Process Coordination
and
Ubiquitous Computing

Process Coordination

and

Ubiquitous Computing

Edited by
Dan C. Marinescu and Craig Lee

CRC PRESS

Boca Raton London New York Washington, D.C.

Cover art is by George Dima, concertmaster of the Bucharest Philharmonic. The title of the drawing is *The Squanderer* (*Risipitorul* in Romanian).

Library of Congress Cataloging-in-Publication Data

Catalog record is available from the Library of Congress

Visit the CRC Press Web site at www.crcpress.com

© 2003 by CRC Press LLC

No claim to original U.S. Government works
International Standard Book Number 0-8493-1470-4
Printed in the United States of America 1 2 3 4 5 6 7 8 9 0
Printed on acid-free paper

Preface

This book is a collection of articles on topics related to process coordination and ubiquitous computing presented at a mini-workshop held on the campus of the University of Central Florida in December 2001.

Process coordination and workflow management have been a topic of considerable interest to the business community for some time. Automation of business processes, the control of factory floor assembly processes, and office automation have contributed to increased productivity and are partially responsible for the positive economical developments we have witnessed during the past decades.

With the advent of computational, data, and service grids, the computer science community is paying increasingly more attention to the theoretical as well as practical aspects of coordination. A grid environment favors resource sharing, as well as intricate collaborative efforts.

A grid is an open system, a collection of autonomous administrative domains. The successful completion of any task in a grid environment requires careful coordination and is subject to multiple optimization constrains; entities with different and possibly conflicting long term objectives must cooperate with one another.

The papers presented at the workshop can be broadly classified in three groups. The first group is devoted to coordination models, the second to coordination on a grid, and the third to intelligent coordination and ubiquitous computing.

The coordination models section opens with a paper by Davide Rossi, *Space-based Coordination and Open Distributed Systems*. The paper advocates well thought out trade-offs between the restrictions imposed by a coordination architecture and the latency, protection, reachability, and other guarantees it offers.

Gian Pietro Picco, Amy Murphy, and Gruia-Catalin Roman propose global virtual data structures (GVDS), an abstraction facilitating the development of application-friendly distributed systems. Lime, PeerWare, XMiddle, incarnations of the GVDS, are reviewed.

In his paper *Models of Coordination and Web-based Systems*, Robert Tolksdorf presents a comparative analysis of several coordination models based upon a set of relevant criteria including stability, awareness, autonomy, and orthogonality. He then describes a Linda-like coordination media that uses XML documents.

Chuang Lin and Dan Marinescu introduces temporal logic models in their paper *Temporal Logic Coordination Models*. The authors argue that workflow analysis and verification is incomplete without a temporal analysis; the paper goes on to introduce Petri Net models for basic workflow patterns.

The paper *A Coordination Model for Secure Collaboration* by Anand Tripathi and co-workers concludes this section. The authors identify the requirements for a secure collaboration in a dynamic environment and present a role-based model for specification of these requirements.

The grid coordination section starts with a paper by Craig Lee and Scott Michel devoted to content-based routing to support events, coordination and topology-aware communication in a grid environment. The author emphasizes the importance of content-based routing, analyzes its complexity and capabilities, and discusses possible implementations. He concludes that topology-aware applications can be more closely coupled and better tolerate the heterogeneous latencies inherent in wide area networks.

The second paper, authored by Dan Marinescu and co-workers, proposes a model for coordination and scheduling on a computational grid based upon the concept of Kolmogorov complexity. The model supports a qualitative analysis of the effects of crossing the boundaries of administrative domains and of the granularity of resource allocation.

The paper co-authored by Alexandre Denis, Christian Perez, Thierry Priol, and Andre Ribes, *Programming the Grid with Distributed Objects* investigates environments for high performance computing capable to exploit Wide Area Network technologies for maximum performance; the authors propose to extend the CORBA architecture with parallel CORBA objects. The results presented in the paper show that the high overhead generally attributed to CORBA is not intrinsic to the architecture but to its implementation.

Frederica Darema contributes an abstract of her presentation on *Dynamic Data Driven Application Systems*. Such systems will be able to dynamically discover and ingest new data from other systems, and allow computational systems with sensors and acuators to be tightly integrated with physical systems thereby creating complex aggregate systems. These systems offer new exciting possibilities and pose new challenges for metadata discovery and the speed and scope interaction.

The last paper of this section *Towards Coordinated Work on the Grid: Data Sharing Through Virtual File Systems* is co-authored by Renato Figueiredo, Nirav Kapadia, and Jose Fortes; it describes a virtual file system supporting sharing across multiple administrative domains. The system uses proxies to mediate transactions among standard Network File Systems and achieves very competitive levels of performance.

The third section on intelligent coordination and ubiquitous computing opens with a paper due to Jeff Bradshaw and his collaborators, a group credited with pioneering work in the area of software agents. The authors argue that *terraforming cyberspace* is a necessity because "cyberspace is a lonely, dangerous, and relatively impoverished place for software agents" and, at the same time, it is possible because "the basic infrastructure with which we can begin the terraforming effort is becoming more available". Social, legal, and life support services are an inherent part of the emerging agent societies in the vision presented in this paper which goes on to describe life support services in NOMADS and legal and social services in KAoS, two of the systems developed by the authors.

Andrea Omicini dwells upon the role of the environment in the context of software agents and introduces the notion of agent coordination context to model and shape the space of agent interactions. The paper presents several examples of agent coordination contexts including the *control room metaphor* and TuCSoN system.

Reflective middeware for ubiquitous computing is the subject of the paper of Koushik Sen and Gul Agha. The middleware provides a representation of its different components to the applications running on top of it; the applications can *inspect* this representation and modify it. The paper describes a host of problems that need to be addressed to fully integrate computing into the physical world.

Do we need to revisit IP routing in view of renewed concerns regarding reliability, security, scalabilty, and above all Quality of Service (QoS) requirements of the Internet? Erol Gelenbe and his co-workers believe so. The *Cognitive Packet Network (CPN)* architecture they propose is capable to provide answers to key issues facing the networking community. Intelligent routing is achieved in CPN using "smart packets"; CPN takes into consideration performance, reliability, and QoS guarantees when routing decisions are made.

Physical and virtual mobility are quintessential elements of the new environments discussed in this book. The paper by Philippe Jacquet provides an overview of the IRULAN project at INRIA-Paris. The project is focused on a mobile ad hoc network in an urban environment.

Data compression is a major concern for multimedia applications in a ubiquitous computing environment and this is precisely the subject of the last paper in this section co-authored by Sacha Zyto, Ananth Grama, and Wojciech Szpankowski. The paper presents matrix transformations for image and video compression.

Finally, Dr. Stephano Trumpy presents the perspective of ICANN (Internet Corporation for Assigned Names and Numbers) on security for the Internet. The Internet and the World Wide Web have become an integral part of the global infrastructure for science, engineering, commerce, and culture. Protecting this vital resource is an essential aspect of ubiquitous computing.

Yongchang Ji had a major contribution to the success of the workshop and the preparation of these proceedings. The organizers of the workshop enjoyed the full support of the administration of the University of Central Florida and of the staff of the Computer Science Department.

Dan C. Marinescu
Craig Lee

Contents

ix

x

Part I - Coordination Models

1

Space-based Coordination and Open Distributed Systems

Davide Rossi

Abstract

Space-based coordination systems (the ones based on shared dataspaces, usually inspired by the Linda language) are widely acknowledged as effective tools to solve concurrent programming problems. By defining simple interactions between distributed components (typically in the context of parallel programs) via access to a shared dataspaces, most concurrent problems find intuitive, elegant solutions. This feature, along with the inherent spatial an temporal decoupling between the components involved in the computation, has attracted many computer scientists interested in applying these techniques in the context of open distributed systems (from the ones running on small networks to the ones running in Internet-scale networks). While the inherent simplicity and elegance of the coordination-based systems is easily preserved, the set of problems arising in open, large-scale, systems (potential network partitions, host failures, large delays, unavailable a-priori knowledge of all the components involved in a computation and so on) needs to be addressed by specific technological solutions (time-out awareness, transactions, and others). Integrating these solutions with a space-based coordination language is, however, never as smooth as desired. Many issues have to be taken into account and the introduced modifications can affect the semantics itself of a coordination language. In this paper we consider the most widespread adaptations to space-based coordination systems to open distributed systems, we analyze the pros and cons of each solution, we present alternative techniques based on novel concepts (such as mobility) and we identify the classes of application to which the coordination systems best adapt to.

1.1 INTRODUCTION

Space-based coordination languages can solve most concurrent programming issues (synchronization and data exchange) with elegance and simplicity. Moreover it has been shown that they can do that with good performances. The use of a shared datas-pace as the coordination medium uncouples the coordinating components in both space and time: a component can produce data independently of the existence of the consumer and can terminate its execution before data is actually accessed; moreover, since the components do not have to be in the same place to interoperate, we can abstract from locality issues. This is why space-based coordination languages are moving from their initial environments (the environments of Linda: parallel comput-ers and workstation clusters) to open systems. When we say a system is open we mean open with respect to several issues. We refer to the fact that the spatial extension of the system is not subject to an a-priori limitation (ranging from a LAN to the Internet). We also refer to the fact that the software can be extended with new services and new components that join the system at run time and the distributed components can be hosted in heterogeneous hardware or platforms (operating systems and middlewares). Or even to the fact that the system can interact at different times with different users.

Open systems should be flexible, adaptable, interoperable. It is easy to see that space-based coordination languages have much to offer in this context. They provide a platform neutral solution to handle synchronization and data exchange between distributed components giving away temporal decoupling support for free. But using space-based coordination languages in open systems is not as straightforward as we would like. As we are going to show is almost impossible to use these languages regardless of the context. A popular adage says: "when the hammer is the only tool every problem looks like a nail"; when people get used to the elegance and simplicity of space-based coordination languages, they plan to use them in every context but, as we will show, most of the times the correct software architecture is needed to use space-based coordination systems effectively.

To better understand where the problem is, let us give a look to the properties that holds in a closed systems and that are unavailable in open ones (most of the times we will assume the Internet as the network infrastructure, both because this is often a worst-case scenario and because of its general interest). First of all in a open systems we miss a-priori knowledge about the users that interact with the system and about the components that are (or will be) part of the software. As we will see this affects performances, security and resource usage. Another property me miss is network reachability and assumptions on delays. As we all know in a asynchronous system there is no way to distinguish between a faulty node and a very slow one [1], but in a LAN context we can make some assumptions that we cannot make in the Internet. The same applies to reachability: within a LAN we can often assume that no partition can occur. Of course the same does not hold in the Internet. Another problem we have

4

when we don't know in advance the components involved in a computation is the space usage. How does one component know this data won't collide with data produced by another component for different purposes? And more: how does one component know the data it is producing will be ever consumed by another component? Maybe the consumer failed, or maybe it won't ever be run. This data will stay forever in a space and is useless. And what a consumer component when the producer is dead or is unable to reach the shared dataspace? Should it be stuck forever waiting for data that will never be produced?

As you can see we have a lot of troubles here. And none of this trouble applies to closed systems; they are "new" and we have to find solutions for them

In the next sections of this article we will see the technological solutions that can be applied to (partially) solve these troubles. Most of these solutions are well known since they are inherited from other technologies deployed in open systems, as we will see, however, not every promising solution fully hits the target, in the forthcoming discussion we will try to point out when this is the case. Section 1.3 discuss the problems related to security and the possible solutions (encryption and access control). Section 1.2 introduces the concept of multiple dataspaces and its variation (hierarchically structured, flat, etc...). Section 1.4 deals with the problem of memory resource usage (when unuseful data clutters a dataspace). Section 1.5 discuss synchronization problems that arise when we a component tries to synchronize with a very slow (or even crashed) counterpart and how to deal with transactional access and its impact on the semantic of a coordination system. Section 1.6 concludes the paper

1.2 NEED FOR STRUCTURE

In a unique shared space every one can hear you scream. Having a unique, flat space can lead to several problems: first of all (as we will see in the next section) there are security issues. If we use the shared space to exchange sensitive data we have to implement a mechanism so that only trusted entities can access these data. Moreover two components might use the same data structure as other components to share completely different informations (and this will lead for sure to the complete misbehaviour of both components sets involved). Even more, if the remote shared space is not distributed (i.e. implemented as a single server) having a unique space means having a single point of failure and a potential bottleneck. This lead to structuring the shared data space in multiple spaces. These space can be flat and unrelated or they can be nested (logically and/or physically). Each single space can be itself distributed of centralized. Well-known spaces can be used by the components to exchange references (or addresses) to different (possibly dynamically created) spaces that are used for specific coordination tasks.

5

1.3 SECURITY

When the central point of a system is a shared dataspace security is obviously a concern. Putting sensible data when everyone can access them is surely a bad idea. Furthermore a malicious entity can break an application by simply removing a data used to coordinate a set of components, even if it is not interested in the contents of the data removed. We have also to deal with the possible attacks from malicious entities that could sniff our traffic with a remote dataspace. The solutions here are quite straightforward: encryption, authentication, access control lists and capability-based systems can be used to overcome the problem. The lack of security has been an argument often used by people opposing to the shared dataspace model. The truth is that security is a problem here as in just any other system in which information exchange takes place. By using the techniques discussed above the same identical degree of security can be reached for shared space-based and message-based systems. Even more we could claim that the use of message-based systems can lead to dangerous confidence; in a shared dataspace system it is much more clear that you need to take care about security, which is a good thing.

1.4 RESOURCE USAGE

When a priori knowledge about the components involved in a coordination task is missing it is simply not possible to know if some data put in a shared dataspace will ever be used or will simply remain there useless, wasting resources (typically memory, but even processing power since the amount of data in a dataspace can affect the computation needed to check for matching items during associative access). Typical approaches inherited from garbage collection techniques adopted in programming languages cannot be applied in this context. Some works (like [3]) present an automatic garbage collection mechanism but only at a space level (the garbage collector claims unused spaces as a whole, it cannot claim unused data in the spaces) and only when a very specific relationship holds among the spaces themselves and only when the space are accessed via unforgeable references (as you can see this is quite restrictive, nevertheless if you can deal with these constraints the system becomes interesting).

In more generic systems there is no way to optimize resource usage the safe way. Once again we have to make assumptions. This leads to the use of time-to-live information associated to the data in the dataspace. When the time-to-live is over the data can be removed. This kind of data is ofter referred to as transient data. Single implementations may vary in the details but the overall concept is the one presented here. As you may see this solution is far from optimal, nevertheless it is the only viable one most of the times. Once again it is clear that many of the solutions to the problems of coordination systems in open environments are sub-optimal.

1.5 SYNCHRONIZATION AND ATOMICITY

A failure in one of the components involved in a coordination task usually has influence on all the other components as well. For simplicity we will consider only two patterns (failures in other coordination patterns have similar consequences). First we analyze the case in which a producer component fails while one (or more) consumer components are waiting for produced items. In this case the consumer component might block forever, and that, of course, is unacceptable. A first approach to this problem is to interrupt the blocking operation after a timeout. Of course this is not a clean solution: the producer might simply be slow. But as already pointed out we cannot distinguish from crashed to slow components so the clean solution simply does not exist. We can say that timeouts are simply a pragmatical approach to a complex problem. But even if we could assume that all slow components really have crashed, timeouts cannot solve all of our problems, as we will see introducing the second coordination pattern we analyze in this section. Consider an item in a dataspace used as a shared counter (an usual way to implement total ordering on a set of items produced by different components or to implement unique IDs). Using the standard Linda-like operators each component interested in getting the current value of the counter and incrementing it, takes the current counter from the shared dataspace, increments it, and puts it back into the space. The sensible point here is what happens if the component crashes (or becomes unable to access the shared space again) after it gets the counter but before putting it back. In general this kind of trouble happens whenever a set of operation has to be performed as a whole in order to preserve the integrity of data in the space (and this includes coordination tasks involving several components). A classic solution is to implement a transaction mechanism. The set of operations to be performed atomically is included in a transaction. If there is a failure during the transaction the shared dataspace is rolled back to the previous state. This is a very well known technology and it is used in several shared dataspace system. But there is a price to pay. To correctly implement the usual ACID properties (Atomicity: the entire sequence of actions must be either completed or aborted; Consistency: the transaction takes the resources from one consistent state to another; Isolation: a transaction's effect is not visible to other transactions until the transaction is committed, Durability: changes made by the committed transaction are permanent and must survive system failure) a lot of locking and serialization is required in the code of the dataspace engine (see [5]), leading to a huge overhead and a to very poor scalability. But transactions also can have subtle impact on the semantics of the systems. Many dataspace-based coordination systems offer predicative operations, i.e. nonblocking input operations that return a false boolean value if there in no element matching the given pattern and true otherwise. In the presence of a transaction a predicative operation may become blocking, depending on the semantic used for these kind of operations. Suppose, in fact, that the semantic is to return a false value only if there is no item matching the argument that has been produced before the operation (by before we refer to the usual happens-before logical relationship in a asynchronous system). This is a reasonable semantic, probably the most reasonable one. Suppose now that an item matching the argument has been produced but it inside a transaction.

7

To respect the semantic and the ACID properties we have to block until the transaction is either committed or aborted. Note that the overall time-semantic of the system with respect to the happens-before relation is preserved, nevertheless blocking predicative operations can be troublesome in interactive applications.

Even more subtle interferences between the semantic of the system and transactions exist when the system uses transient data.

Very recently, to overcome these troubles, transactions with relaxed ACID properties or optimistic transactions have been proposed but relevant works on this subjects still have to appear.

A different approach to support atomic operations is based on mobile code. By bulk-moving the set of operations to be performed atomically to the host, or the LAN, where the dataspace is hosted we can overcome part of the problem (see [7]). Of course this holds just when the operations involve a single component (like in the case of the shared counter example given before), when the coordination task to execute atomically involves more components the things are much more complex. Another approach is to download to the space engine code to implement new primitives that can perform several operations at once (as in [9]). In both cases resource usage (to run the mobile code) is a relevant issue, and to enforce a limit on resource usage can be technically troublesome.

1.6 CONCLUSIONS

As we have seen, even when deploying a full set of technological solutions, the usage of a space-based coordination system, as-is, in a open environment is, at least, troublesome. To deploy with success this kind of systems into open environments we need the right software architecture. For example mobile agents systems in which coordination tasks among agents take place only (or preferably) when they are co-located (see [10]). Or collaborative distributed work systems in which coordination tasks are used to implement workflow-based engines that run in a local environment [11]. Even agent-based Web-accessible architecture have been proposed (see [12]) in which coordination is used in a limited environment. All in all the key point is to identify architectures in which the components involved in coordination tasks are run in environments that can give a certain amount of guarantees (about latencies, delays, reachability, protection, resource availability and so on) without imposing too much restrictive constraints. As of today these have been the only successful implementation of coordination systems in open environments. As per general usage, RPC-based and distributed events-based middlewares are much more successful (even if most RPC-based systems present an amount of troubles in open environments that is no less than the amount of troubles space-based coordination systems have). Surely event-based middleware looks more promising in open environments and some coordination systems implement some kind of event-notification mechanism on top of the generative communication paradigm but it is unclear wether they can inherit the benefits of pure event-based systems.

REFERENCES

1. K. M. Chandy and J. Misra, *How processes learn.* In Distributed Compuring, 1986.

2. T. Kielmann, *Designing a Coordination Model for Open Systems.* In Proc. 1st Int. Conf. on Coordination Models and Languages, Springer-Verlag, 1996.

3. R. Menezes, *Experience with memory management in open Linda systems.* In Selected Areas in Cryptography, 2001.

4. R. Menezes, R. Tolksdorf and A.M. Wood, *Scalability in Linda-like Coordination Systems.* In Coordination of Internet Agents: Models, Technologies, and Applications, Springer-Verlag, 2001.

5. N. Busi and G. Zavattaro, *On the Serializability of Transactions in JavaSpaces.* In Proc. International Workshop on Concurrency and Coordination (CONCOORD'01), Elsevier, 2001.

6. N. Busi and G. Zavattaro, *On the Serializability of Transactions in Shared Dataspaces with Temporary Data.* In Proc. 2002 ACM Symposium on Applied Computing (SAC'02), ACM Press, 2001.

7. A. Rowstron, *Mobile Co-ordination: Providing fault tolerance in tuple space based co-ordination languages.* In Proc. Coordination Languages and Models (Coordination'99), Springer-Verlag, 2001.

8. A. Rowstron, *Using agent wills to provide fault-tolerance in distributed shared memory systems.* In Proc. 8th EUROMICRO Workshop on Parallel and Distributed Processing, IEEE Press, 2001.

9. G. Cabri, L. Leonardi and F. Zambonelli, *MARS: A Programmable Coordination Architecture for Mobile Agents.* IEEE Internet Computing, July-August 2000.

10. P. Ciancarini, A. Giovannini and D. Rossi, *Mobility and Coordination for Distributed Java Applications,* In Advances in Distributed Systems, Springer-Verlag, 2001.

11. D. Rossi, *The X-Folders Project.* In Proc. European Research Seminar in Advanced Distributed Systems (ERSADS 2001), 2001.

12. P. Ciancarini, R. Tolksdorf, F. Vitali, D. Rossi and A. Knoche, *Coordinating Multiagent Applications on the WWW: a Reference Architecture.* IEEE Transactions on Software Engineering, 24(5), 1998.

Author(s) affiliation:

- **Davide Rossi**

  Dipartimento di Scienze dell'Informazione
  Università degli Studi di Bologna
  Bologna, Italy
  Email: rossi@cs.unibo.it

2

On Global Virtual Data Structures

Gian Pietro Picco
Amy L. Murphy
Gruia-Catalin Roman

Abstract

In distributed computing, global information is rarely available and most actions are carried out locally. However, when proving system properties we frequently turn to defining abstractions of the global state, and when programming we often find it convenient to think of a distributed system as a global centralized resource. In this paper we build upon this observation and propose the notion of *global virtual data structures* as a model for building a new generation coordination models and middleware that allows programmers to think of local actions as having a global impact and places upon the underlying system the burden of preserving this appearance. The model itself is inherently peer-to-peer, lending itself toward applications which are largely decentralized, and built out of autonomous components.

2.1 INTRODUCTION

Distributed computing is no longer the exception, but rather the rule. More often than not, the loss of network connectivity severely cripples if not completely destroys the ability to continue processing. As networks continue to increase in size, pervasiveness, and complexity, it becomes more difficult to both conceptualize the distributed system and understand the consequences of interacting with various parts. Further complicating these issues is the existence of multiple users, simultaneously interacting with the same system, making changes and viewing the changes made by others.

Moreover, distributed computing is increasingly moving from a rather static and controlled architecture, stigmatized by the client-server paradigm, towards more autonomous, decentralized, and dynamic settings. For instance, Two environments exposing such dynamicity are those defined by peer-to-peer networks and by mobility. In peer-to-peer systems, the members of the community are constantly changing, and each peer is individually responsible for the data which they are sharing with the rest of the system and for the propagation of queries operating on the data. Mobility adds an additional level of complexity by allowing hosts to move through space, altering the shape of the underlying communication framework along with the changes in connected components as hosts move in and out of wireless communication range. Another level of dynamicity comes from the applications themselves which must change to adapt to the changing topologies and available data. Such applications may be formed out of logically mobile components which can relocate to better utilize local resources or perform load balancing. All of these dynamic aspects inherent to these environments increase the effort in application development.

In this work we present an abstract model which, unlike other models, does not identify a single view of distributed environments, rather it lays a foundation for building a broad range of models. Our work is grounded in a study of coordination models for distributed computing, which separates the computation, or the task-specific programming, from the communication, or the interaction among processes. Distributed coordination models consider the need to take local decisions while still conceptualizing the effects of these actions at the global level. Typical coordination models attempt to simplify the programming task by hiding the distribution of the environment, providing operations to users which have the same effect regardless of the location of the information being acted upon. Our work distinguishes itself by seeking not to hide the inherent distribution, rather to take advantage of the distribution, enabling a richer set of operations and more flexible user interactions.

Despite this added complexity, our driving design strategy is summarized by the desire to coordinate distributed applications by thinking globally while acting locally. Put simply, the user works with an abstract global view of the system, and while all operations are issued in the same manner as local operations, their effect is defined with respect to this global view. An important feature of our abstract model is its applicability to a wide range of distributed environments (from large, wired, wide area networks to wireless, ad hoc environments) and a diversity of application domains (from data-oriented to dynamic, configuration-centered systems).

Our model, termed *global virtual data structures* (GVDS), is essentially a coordination model. For the programmer of an individual software component, GVDS manifests itself in the form of a common data structure accessible via a standard set of operations. This data structure, however, is not real but only a reflection of the global state of the system further constrained by its current configuration, i.e., by a set of rules that establish how much of the global state is actually accessible at each point in time and how much accuracy this representation carries with it. Of particular interest is the ability to define the projection of accessible data with respect to the component issuing the operations. In other words, it is possible for two connected components to have different views of the data structure depending on user-defined constraints. The choice of abstract representation and the relation between the global state and its virtual local view are the differentiating factors among different instantiations of the GVDS concept.

The paper is structured as follows. Section 2.2 aims at defining precisely, albeit informally, the notion of GVDS. Section 2.3 presents some existing models and systems that incarnate this concept. Section 2.4 highlights the main assets of GVDS. Section 2.5 discusses the design challenges and alternatives of a system that adopts a GVDS perspective and highlights the research opportunities involved. Section 2.6 elaborates further on the notion of GVDS, discussing its relationship with other approaches and systems, and identifying research challenges. Finally, Section 2.7 ends the paper with some concluding remarks.

2.2 CONCEPT DEFINITION

The problem of providing distributed access to information through a programmable medium can be regarded as a coordination problem. Coordination is a programming paradigm that seeks to separate the definition of components from the mechanics of interaction. In traditional models of concurrency, processes communicate with each other via messages or shared variables. The code of each component is explicit about the use of communication primitives and the components are very much aware of each other's presence. Actually, communication fails when one of the parties is missing. By contrast, coordination approaches promote a certain level of decoupling among processes and the code is usually less explicit about mechanics of the interactions. Ideally, the interactions are defined totally outside the component's code.

Coordination is typically achieved through implicit communication among the components. Linda [9] is generally credited with bringing coordination to the attention of the programming community, but other models (e.g., those based on CHAM [4]) take a similar perspective. Agents exchange information indirectly through manipulation of a shared data structure, with no need for the parties involved to know about each other. Usually, additional decoupling between the communicating parties is provided by the fact that some degree of persistency of the data structure is allowed, hence the communicating parties are freed from the constraint of being available for communication at the same time and/or at a well-known location. In traditional coordination models the data structure that represents the coordination medium is assumed to be *indivisible, persistent, and globally available* to each of the coordinated component.

Clearly, these assumptions clash with the highly dynamic scenario we are targeting in this paper. Models based on a GVDS retain a coordination perspective, fostering

13

decoupling of behavior and interaction, but reconcile it with the requirement of dynamicity by lifting precisely the aforementioned assumptions.

In a GVDS, the data structure serving as a coordination media is distributed according to some well-defined rule among the coordinated components. Each component is then associated with a fragment of the overall data structure, that can be manipulated through the operations defined on it. Distribution may involve the schema of the data structure (e.g., in the case of a tree where subtrees are assigned to different components) or the content of the data structure (e.g., different subsets of a set). Hence, the coordination data space is no longer indivisible, rather it is, by definition, made of separate parts.

Nevertheless, the coordination space is still perceived as a single entity, since the local data owned by the components are transiently and dynamically shared under a single, global data structure—the GVDS. However, sharing occurs only among those components that are mutually reachable. By ignoring connectivity constraints, combining the data of all these separate parts forms a single, global data structure. However, because of constraints of the dynamic environment, not all parts are simultaneously available. Therefore, this global data structure is *virtual*, meaning no single component can view it as a whole, but all components know of its existence. The part of the GVDS which is concretely visible to an individual component is defined by the contents of the data owned by the individual components that are reachable. Typically, physical connectivity among components is assumed as the rule determining the maximal local view. While this a reasonable and useful assumption, and only a precondition for more elaborate rules. For instance, components belonging to the same network partition, determined according to physical connectivity in a mobile ad hoc network, may actually be considered belonging to two different partitions under a GVDS where partitions are determined according to security domains. In any case, the data structure exploited for coordination is no longer globally available, rather it is available only within a partition. Moreover, it is not persistent either, since partitions are in general allowed to change dynamically.

Interestingly, access to this shared data structure is granted through the operations defined on the local data structure. Hence, from the point of view of components, the distinction between local and global data becomes blurred, and a single way of interacting with the rest of the system is provided. Moreover, it is worth noting that the notion of GVDS defines a model of coordination that is intrinsically peer-to-peer and component-centric. In other models, the data structure that holds the information relevant to coordination is totally external to the coordinated components, whose contributed data are essentially indistinguishable. Instead, the notion of GVDS clearly separates not only the behavior of the component from the interaction, but also the data provided by the component from that present in the rest of the system.

The notion of GVDS we just defined informally can then be characterized by the following fundamental properties. A GVDS is:

- *Distributed.* Parts of the data structure are owned by different components, that store them locally.

14

- *Constructive.* A component's local view of the GVDS is formed through some combination of the data which is on reachable components.

- *Uniformly accessed.* All the components access the GVDS through the same set of operations.

- *Parochial.* Operations appear to be performed locally, even though they may access distant data or have global effects.

These properties are concisely summarized by the name "global virtual data structure", meaning that the union of the system-wide data forms a global data structure. Operations have potentially global effect, but no single component can access the entire structure, making the data structure virtual from the perspective of the component.

The notion of GVDS represents a meta-model for coordination. It provides a conceptual framework that can be leveraged off to define coordination models addressing specific needs. Specific instantiations of the GVDS concept may differ in several respects. For instance, not only the data structure of choice and the corresponding operations may be different, but also the rules for dividing the data structure and merging it back may vary. Appropriate selection among these and other alternatives are the key to a successful use of the GVDS concept. Moreover, the resolution of the design tradeoffs involved are likely to be the source of a new way of looking at old research problems, if not a source of new ideas.

Before examining these design alternatives and research challenges, we now turn our attention to some existing models and systems based on the notion of GVDS, to show some possible instantiations of this meta-model. We will come back to the issue of how to design a coordination model that exploits GVDS in Section 2.5.

2.3 GVDS INCARNATIONS

In this section, we describe a few existing systems that are based on the notion of GVDS. This description serves the purpose of making our discussion of GVDS more concrete by looking at existing models and systems, and of exemplifying some the choices that can be made when defining a specific instantiation of the GVDS meta-model.

2.3.1 Lime

LIME (Linda in a Mobile Environment) [14, 12, 15] is a coordination model and middleware designed to support the development of applications involving physical and logical mobility. Many of the ideas put forth in this paper are inspired by our own experience in designing and implementing LIME.

In LIME, the data structure chosen for the GVDS is a Linda-like tuple space. Tuples can be inserted in the tuple space using out, read using rd, and withdrawn using in. Tuple spaces are permanently associated to the *agents* in the system, which can be thought as components with a thread of control. Agents can be mobile, and can move across *hosts* carrying along their tuple space, called the *interface tuple space*, during migration. Hosts can be themselves mobile, and yet retain communication through wireless links.

The transient sharing provided by the GVDS perspective becomes key in rejoining physical and logical mobility under a single coordination approach. As shown in Figure 2.1, sharing occurs at two levels, based on different notions of connectivity. When two or more mobile agents are co-located on a given host, their interface tuple spaces become transiently shared. The resulting tuple space is called a *host-level tuple space*. Similarly, when two or more mobile hosts are in communication range, their host-level tuple spaces become transiently shared under a single *federated tuple space*.

The meaning of Linda operations is redefined in LIME to take into account transient sharing. Operations no longer manipulate a globally available and persistent tuple space, rather they are defined on an agent's interface tuple space. However, because of transient sharing, they effectively operate on the whole federated tuple space, provided that other agents are connected. Hence, the idea of transiently shared tuple spaces reduces the details of distribution and mobility to changes in what is perceived as a local tuple space. This view is powerful as it relieves the designer from specifically addressing configuration changes, but sometimes applications may need to address explicitly the distributed nature of data for performance or optimization reasons. For this reason, LIME extends Linda operations with location parameters, expressed in terms of agent or host identifiers, that restrict the scope of operations to a given projection of the transiently shared tuple space.

The out$[\lambda](t)$ operation extends out by allowing the programmer to specify that the tuple t must be placed within the tuple space of agent λ. "Misplaced" tuples, i.e., tuples destined to an agent λ that is currently not connected, are kept in the caller's tuple space until λ becomes part of the system. This way, the default policy of keeping the tuple in the caller's context until it is withdrawn can be overridden, and more elaborate schemes for transient communication can be developed. Location parameters are also used to annotate the other operations to allow access to a slice of the current context. For instance, rd$[\omega, \lambda](p)$ looks for tuples matching p that are currently located at ω but destined to λ.

LIME extends Linda by introducing also a notion of *reaction*, motivated by the fact that, in the dynamic environment defined by mobility, reacting to changes is a big fraction of application design. A reaction $R(s, p)$ is defined by a code fragment s specifying the actions to be performed when a tuple matching the pattern p is found in the tuple space. Details about the semantics of reactions can be found in [14, 12]. Here, it suffices to note that two kinds of reactions are provided. Strong reactions couple in a single atomic step the detection of a tuple matching p and the execution of s. Instead, weak reactions decouple the two by allowing execution to take place eventually after detection. Strong reactions are useful to react locally to a host, while weak reactions are suitable for use across hosts, and hence on the federated tuple space.

16

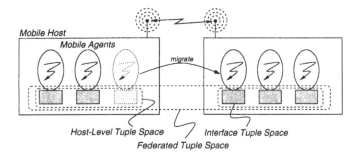

Fig. 2.1 Combining the tuple spaces of the individual components (mobile agents) to form the federated tuple space.

The LIME model has been embodied in a middleware available as open source at lime.sourceforge.net). We have successfully built a variety of applications with LIME [15], ranging from a data-centered multi-player jigsaw assembly program to a system-configuration centered spatial game involving teams of players moving through physical space and interacting when connectivity is enabled. LIME is also being exploited in an industrial collaboration in the context of automotive applications.

2.3.2 PeerWare

PEERWARE [8], under development at Politecnico di Milano, is a middleware that aims at providing core support for systems exploiting a peer-to-peer architecture, hence including mobile systems. While this goal is very similar to the one of LIME, the two systems differ in the choices made while instantiating in their own model the GVDS meta-model.

The data structure chosen in PEERWARE is a tree or, more precisely, a forest of trees. Each component of the system, called *peer* in PEERWARE, hosts several trees with distinct roots. The *nodes* of each tree (see Figure 2.2) are essentially containers for *documents* holding the actual data belonging to the application context. When connectivity among peers is established, the GVDS that is dynamically reconstructed is defined as follows. All the *homologous* nodes, i.e., nodes with the same name and holding the same position in trees belonging to different peers, are represented by a node in the GVDS having the same name and position as the concrete ones. The content of such a node in the GVDS is the union of the documents contained in each homologous node. A pictorial representation of this process is provided in Figure 2.3.

PEERWARE distinguishes between the local data structure and the global one, by providing two different sets of operations. The rationale behind this choice is to make the distinction between local and global access, as well as its implications, explicit to the programmer. This is a different approach with respect to LIME, where the GVDS subsumes the local data structures, and syntactic means are used to restrict scope.

17

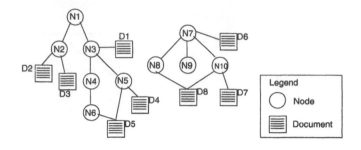

Fig. 2.2 The data structure provided by PEERWARE.

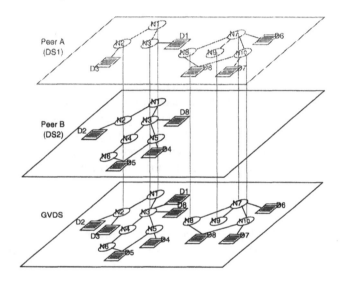

Fig. 2.3 Building the GVDS in PEERWARE.

Hence, a set of operations is defined only on the local data structure. These operations are concerned with local data modification, such as creation and removal of nodes and documents. Notably, PEERWARE does not provide any feature for modifying remote data, while this is available in LIME in the form of out[λ]. In addition, a publish primitive is provided that, similarly to publish-subscribe systems [16], allows the generation of an event occurring on a document or node.

Another set of operations is defined both on the local data structure and the GVDS. The operations allowed on the GVDS are very limited in number, but rather expressive in power. The basic access primitive is execute, which accepts as parameters a set of *filters* and an *action*. The filters are expressions that allow one to restrict the set of documents and nodes (collectively referred to as *items*) the execute operates on. In turn, the action specifies the sequence of operations that must be performed on such items before being returned to the caller. The simplest action is clearly an empty one.

18

In this case, the effect of the action is simply to return the set of items selected by the filters, unchanged. In general, however, the programmer has great flexibility in defining the behavior of actions, thus effectively redefining the semantics of access to the data structure. For instance, actions may process further the filtered items (e.g., by performing additional filtering or changing the items' properties) and may also access again the data structure (e.g., to remove the selected items). Essentially, execute provides a core mechanism to disseminate, among the currently connected peers, a request for local execution of application code accessing the local data structure. The execution of a single action is atomic on each receiving peer (i.e., actions coming from different peers do not interfere), but not across several peers.

Besides execute, PEERWARE provides also a subscribe primitive, whose semantics is similar to that of publish-subscribe systems, but operates on the *state* of the system, rather than on generic application events. Thus, subscribe allows a peer to register interest for events occurring on a subset of the data structure content, as specified by a filter. Finally, the last primitive provided is executeAndSubscribe which, as the name suggests, is essentially a combination of the previous ones. This primitive essentially allows the programmer to "hook on" a given set of data by performing an execute and, in the same atomic step, a subscribe. This allows the programmer to implement schemes with a consistency that is stronger than the one that could be achieved by using execute and subscribe separately, since this latter solution would allow events occurring in between the two operations to be lost.

PEERWARE is currently being exploited as the middleware layer in the MOTION[1] project, funded by the European Union. The goal of the project is to build a framework of teamwork services supporting cooperation among geographically distributed and mobile users in an enterprise-wide infrastructure.

2.3.3 XMiddle

The final example we consider is the XML-based middleware XMIDDLE [11] under development at University College London. The main principles behind this work are simple connectivity rules and powerful, high-level data operations. In XMIDDLE, the data structure is a set of XML documents whose content can individually be viewed as trees. Documents are combined by mounting one tree at a well defined point in another tree, thus essentially embedding one local document inside another or indicating a mount point for a remote document, much the way remote file systems are mounted in operating systems. Figure 2.4 shows a simple example of two hosts, where the mobile host mounts two subtrees of the master. The resulting local data structures are neither symmetric, nor are connections transitively shared.

Two of the more powerful features of XMIDDLE are its support for data replication and offline operations. While connected, the data of two components are kept synchronized by pushing updates along the pairwise connections. When disconnected,

---

[1]MObile Teamwork Infrastructure for Organisations Networking, IST-1999-11400, www.motion.softeco.it.

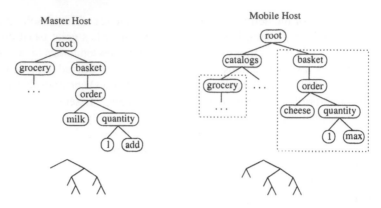

Fig. 2.4 Two disconnected XMIDDLE hosts and their XML tree view of the data. The mobile host mounts the directories for the catalog and the basket (outlined with dotted lines), and makes updates while offline (adding an order). After reconnection of the hosts, the views of the trees are updated, and form the abstract view shown below each XML tree. Note the asymmetry between the master host's data structure and that of the mobile host.

changes are still allowed to the replicated portions of the tree, but may also lead to inconsistencies of the data. These inconsistencies are reconciled upon reconnection using a set of application defined reconciliation rules which are embedded in the XML document itself. In the example of Figure 2.4, the reconciliation rules for orders are to add or take the max of the quantity values. Another option is to leave both conflicting elements and notify the user.

The XMIDDLE system has been successfully used to implement an electronic shopping system where multiple users in the same family can mount the same catalog of items and collectively build a shopping list in a distributed fashion. The replication allows changes in both the catalog and shopping list to propagate among members of the family and the reconciliation rules address any conflicts which may arise from multiple individuals adding the same item to the list.

2.4 THE ASSETS OF A GVDS APPROACH

A natural question to pose at this point is what are the key advantages from a coordination model adopting a GVDS-based perspective. From the illustration of the GVDS concept and the brief overview of the models exploiting it, it appears that in coordination models exploiting the notion of GVDS, the association between the coordination context contributed by a given component and the component itself is made explicit. Hence, the information provided by a given component is no longer dispersed in some coordination "soup" external to the components. Moreover, this association between a component and the data it is willing to share for the sake of coordination models more naturally what is locally available to the component and what instead needs to be looked for in the rest of the system.

Hence, GVDS fosters a coordination style where:

- the resulting model is truly distributed, and the coordination focus is shifted from an external coordination data space back to the components;

- the sharp decoupling between communication and behavior, typical of coordination, is retained: a globally available coordination data space is still available, only its contents are now automatically and dynamically reconfigured according to changes in the system;

- coordination is achieved through local actions that have a global effect.

The conjecture is that these characteristics are going to simplify the task of building—and reasoning about—applications that are built out of autonomous components whose relationships are dynamically and frequently reconfigured. Clearly, a lot about the truth of this latter statement depends on how the notion of GVDS is instantiated into a specific model. The alternatives and tradeoffs available are the subject of the next section.

2.5   FROM THE META-MODEL TO MODELS AND SYSTEMS

As we mentioned earlier, we regard the concept of GVDS defined informally in Section 2.2 as a meta-model of coordination. The quick survey of models and systems we carried out in Section 2.3 seems to suggest that the design space available is rather broad. In this section we examine more closely the space of design alternatives, by setting ourselves free from the details concerned with specific systems. The outcome of this investigation is a preliminary set of issues that we deem relevant in the design of a GVDS-based model and/or system, based on our own experience to date. This list of issues is by no means meant to be exhaustive. As a matter of fact, in defining the content of this section we are essentially defining the agenda for a new area of research on coordination models.

Data structure. Obviously, the choice of the data structure is the first one that needs to be faced. This choice is fundamental in all GVDS-based models, and is usually determined by the overall requirements or design criteria for the model. For instance, the tree structure of PEERWARE naturally suited the requirements of collaborative applications. In XMIDDLE, the choice was driven by the need to merge XML files.

However, the choice of the data structure has an impact that goes well beyond the suitability of the data structure for the particular requirements at hand, affecting non-functional dimensions like performance or security. For instance, LIME exploits a data structure that is essentially flat, although multiple federated tuple spaces can coexist, distinct through names. Instead, PEERWARE may leverage off the hierarchical structure of trees to reduce the amount of traffic generated by queries. Similar considerations apply if security is at stake. A tree structure naturally lends to a security architecture where each tree node is a protection domain, potentially inheriting (and modifying) access rights from the containing tree. Hence, different parts of the

21

same data structure could be made visible to different components. In LIME, protection domains would need to be mapped on tuple spaces, thus providing a single, all-or-nothing mechanism for access control.

Operations. A data structure is intimately connected with the operations that are allowed on it, which constitute another fundamental decision in defining a GVDS-based model. Query operations, allowing to retrieve one or more elements from the data structure, are often a natural choice, in that they simplify the task of querying at once the local data associated with a potentially large number of distributed components. Nevertheless, query operations (like an rd or in in a tuple space, or the selection of an element in a matrix) define a "pull" approach, where a component must proactively retrieve the data it needs. In scenarios where the system configuration is continuously changing, and where the benefits of a GVDS approach are more evident, it may be preferable to complement, or even substitute, proactive operations with reactive operations, thus fostering a "push" approach where the burden of delivering the relevant data to the component is placed on the system. This observation is actually what motivates the presence of reactions in LIME and of publish-subscribe features in PEERWARE. Our experience in building applications with LIME [15] shows how the reactive features of LIME are used much more than the proactive ones in the development of applications for mobile computing.

Clearly, other approaches are possible, like primitives to navigate through graph-like data structures distributed across multiple sites, similarly to what is provided by distributed database management systems supporting the network data model or the object-oriented data model.

Local vs. global. Another relevant design issue about operations is whether the distinction between local and global operations should be apparent from the model (and hence the system) or not. In a "pure" GVDS, there is no distinction between what is local and what is global. Operations are simply performed through a handle to the data structure, and their effect is determined by the input parameters and by the current configuration of the system. However, such a radical location-transparent approach may have significant drawbacks. As noted in literature [17], masking the distinction between local and remote may have a detrimental effect on the overall performance of the system. In LIME, the design tradeoff between location-aware and location-transparent data access is tackled by annotating operations with location parameters that allow the programmer to restrict the scope of queries. In PEERWARE, the distinction is made even more evident by providing two separate interfaces for accessing the local data structure and the global one. The problem is not raised only by operations for data retrieval. For instance, the ability to ship a tuple to a different agent provided by the LIME out[λ] forces the system to buffer tuples and ship them around as soon as the target component shows up. This feature complicates slightly the implementation of the system. At the same time, it is one of the most expressive features of LIME, raising considerably the programming abstraction level. In PEER-WARE, writing data to a remote component is assumed to be handled externally to the model, thus greatly simplifying the implementation at the price of expressiveness and conceptual uniformity.

Rules for dynamic sharing. The last topic is strongly related to another cornerstone of the design of a GVDS model, namely, the definition of the rules governing transient sharing. Three issues are relevant: the conditions under which transient sharing is enabled, the rules defining the shape of the global data structure resulting from sharing, and the amount of computation that is possibly carried out when a change in the configuration, enabling or disabling the sharing of one or more components, occurs.

The first issue is typically tied to some notion of connectivity. For instance, in LIME two notions of connectivity are exploited. Two mobile agents are "connected" (and hence share their tuple spaces) when they are co-located on a host; two mobile hosts are "connected" when they are in range. Clearly, the two can be combined. Weaker notions of connectivity are also possible, e.g., encompassing the ability to communicate through a sequence of hosts that are not all connected at the same time, along the lines of [7]. XMIDDLE enables transient sharing by considering only pairwise connectivity.

While physical connectivity is an intuitive way to define the condition under which sharing may occur, this notion is more properly regarded as a baseline for more sophisticated triggering rules. For instance, transient sharing may be subject not only to the ability to communicate, but also to the satisfaction of security requirements, e.g., belonging to the same administrative domain. A generalized notion of connectivity can be defined, that extends the physical one with application-dependent conditions.

The second aspect related with the details of transient sharing is the definition of the rules determining the perceived content of the shared structure. Clearly, this issue is intimately tied to the specifics of the data structure. The systems we surveyed in Section 2.3 define the shared data structure either as a union or as a composition of the local ones. Names play a key role in defining how the shared data structure is constructed. Both LIME and PEERWARE enable sharing of data on a per-name basis, but while in LIME names are singletons, in PEERWARE they have a structure reflecting the tree. While this fact may appear as a petty technicality at the model level, allowing structured names may have a significant impact at the implementation level, as we discussed earlier. In contrast, XMIDDLE exploits trees in a different way, by defining the shared tree through composition of the local subtrees.

Another relevant issue is the degree of symmetry of sharing. In LIME and PEERWARE sharing is symmetric, i.e., all the components involved in the transient sharing perceive the same data structure. Instead, in XMIDDLE sharing is asymmetric, since each of the components involved in a pairwise sharing may compose the tree in a different way. In addition, physical or logical constraints may introduce asymmetry in the definition of transient sharing. For instance, unidirectional wireless links may enable only one of the components engaged in a pairwise connection to output data in the data structure belonging to the other, but not vice versa. Analogously, security constraints may allow one of the components to access data from the other, but not vice versa.

The third and last aspect directly related to transient sharing is the amount of computation occurring when sharing is triggered. The dominant tradeoff involved is likely to be the one between expressive power and performance. As we mentioned earlier, in PEERWARE no user-specified computation is triggered upon engagement (although control information about the tree structure needs to be exchanged behind the scenes in the implementation). In LIME, the provision of primitives that allow transparent relocation of tuples forces the model to take into account the reconciliation of the misplaced tuples, i.e., tuples generated when the target component was not yet part of the system. In XMIDDLE, sharing involves replication, to increase data availability during disconnection. However, replication is carried out entirely by the system and is not made explicit at the model level.

Other Issues. This last remark raises a number of interesting questions. For instance, is it possible and reasonable to specify a replication behavior by exploiting the very primitives provided by the model at hand? For instance, preliminary experience with LIME showed that its reactive features enable the creation of a higher-level veneer on top of LIME dealing with replication. The concept could be pushed even further, allowing the programmer to specify several veneers that enhance the system by providing some amount of self-reconfiguration or reflection.

Of course, in designing a GVDS-based model of coordination, a number of other well-known concerns must be tackled, like the choice between synchronous and asynchronous primitives and the degree of atomicity they exhibit. The latter one is likely to be the most critical in defining a successful model. Ideally, operations carried out on the GVDS should have the same semantics as the same operations carried out on a data structure that is local to the caller. In practice, however, this is often not achievable. Striking a good balance between expressiveness and scalability is particularly delicate in this case.

2.6 DISCUSSION

In this section, we elaborate further on the concept of GVDS we put forth in this paper, by highlight some of its relationships with other models and systems, and by identifying open research issues and challenges.

2.6.1 Intellectual Connections

Coordination models have been categorized along the lines of data-oriented or process-oriented [13] depending on whether it is changes in the data which drive the application, or events raised by the components which are the focal point. The GVDS meta-model encompasses both with the main emphasis placed on the data structure and its basic operations. However, as has been noted, reactive constructs similar to the events of a process-driven model are sometime key to application development in dynamic environments and are clearly supported in the GVDS model. As we have revisited the Linda coordination model to form LIME, other coordination models may be extended to the unique environment of GVDS applications by studying them in this new light.

It is interesting to note how the GVDS concept to some extent subsumes more traditional models. For instance, the provision of location-aware primitives may subsume the traditional client-server model as a special case of GVDS. In LIME, operations annotated with location information allow for the retrieval of data from a specific host or agent, in a way that closely resembles a client-server interaction. Also, while a GVDS encompasses a coordination data space that can be very dynamic and characterized by transient sharing, in doing so it includes also scenarios where the configuration of the system is partially, or completely, static. Hence, for instance, the transiently shared tuple space defined in LIME "defaults" to a Linda tuple space in the case where all agents and hosts are stationary and connected. This fact can be exploited in practice in applications that need to have some data held persistently: at deployment time, some of the nodes may be designated to be stationary and hence providing a persistent base for the rest of the system, as they will always be able to share their content with any non-stationary component in range.

Also, some systems that are usually not labeled as GVDS-based exploit an architecture that can be framed under this concept. This is the case of systems supporting peer-to-peer file sharing over the Internet, such as Napster [1], Gnutella [2], or Freenet [3]. In these systems, users share the content of a fraction of their file system with other users. Typically, the content is available only when the user is connected, although caching schemes (such as those exploited by Freenet) allow the creation of distributed replicas. To each user, the set of files present in the repository of connected users looks effectively like a global repository—which can be regarded as a GVDS. The users are allowed to perform queries to search for a given file in this global structure, and requests are automatically routed to the users in the system. In these applications, the scope of queries is typically determined by considerations related to performance rather than expressiveness: queries typically have a time-to-live that effectively restricts their extent.

The notion of GVDS is somehow reminiscent of the notion of *distributed shared memory*, in which multiple processes map the same piece of memory and interact through basic read, write and lock operations. The typical emphasis of these systems is on supporting heterogeneous architectures and multiple languages. Recent research has focused on increasing performance by allowing the application programmer to tune the level of consistency required of the data [18, 6], where increased coherence requirements result in higher overhead as compared to weaker coherence models. In general, distributed shared memory systems assume a well known mechanism to find the server of a memory segment, and further assume connectivity between the client and server for the duration of the interaction. Thus, they do not seamlessly address many of the issues of dynamic environments we describe as typical GVDS scenarios.

Some systems have begun to address mobility as one step toward supporting dynamic environments. For example, the Coda file system [10] allows users to cache files and work offline, performing reconciliation when the user reconnects to the master file server. The Deno distributed object system [5] supports object replication and a distributed voting scheme for managing disconnected object updates. While these systems do support mobility, they do not address all of the issues of a GVDS, and ignore high level coordination operations over the managed data. Still, some of the mechanisms exploited by these systems are likely to inspire the implementation of some features of GVDS-based systems.

2.6.2 Research Challenges

We see GVDS as a meta-model, a conceptual framework guiding the definition of novel coordination models and systems. As we already mentioned, the notion of GVDS we shaped in this paper raises a number of open issues, that can be regarded as defining a novel line of research. They include:

- *What is a good balance to strike among design alternatives?* In Section 2.5 we highlighted some of the dimensions determining the move from the GVDS meta-model into a coordination model, and how these decisions affect the tradeoffs between expressiveness and performance. Clearly, the identification of these tradeoffs is a key to the success of GVDS-based models, and especially systems.

- *What are the implications of GVDS for security?* The architecture implied by a GVDS is intrinsically radically decentralized and open. At least at the model level, no central point of control is assumed, and in addition components may come and go at will. Finally, the characteristics of implicit communication, typical of coordination approaches, make it more difficult to define security policies based on authorization. Hence, it reasonable to ask whether the background of security techniques, e.g., involving access control or authentication, are sufficient to deal with this scenario, or whether new models (e.g., of trust) are needed. Interestingly, the problem as defined is analogous to the problem of ensuring security and trust in a peer-to-peer network.

- *Can the choice of the data structure be separated from the distribution and communication aspects?* Is it reasonable to define a sort of GVDS "schema language", analogous to what is done in DBMS systems, allowing the programmer to define the shape of the distributed data structure, and keep this definition decoupled from the details of distribution? And, if yes, to what extent?

- *Is there a theory of GVDS?* Is it possible to capture the notion of GVDS in a formal framework, possibly separating the data issues (e.g., the "schema" of the data structure) from the distribution and communication issues? And, once this is done, how can this framework be (re)used to refine the meta-model into an actual model?

- *What is the impact of GVDS on formal reasoning and verification?* We hinted earlier at the conjecture that the notion of GVDS, by shifting the focus from the coordination infrastructure back to the coordinated components, provides a more natural way to prove properties about the overall system. However, this

26

is only a conjecture, and needs to be verified on the field. Moreover, what kind of formalism or notation is appropriate to reason about GVDS-based systems?

2.7 CONCLUSIONS

Distributed systems exhibit increasing degrees of decentralization, autonomy, and dynamic reconfiguration. In this paper, we defined the concept of *global virtual data structure*, a novel coordination approach that fosters a component-centered perspective by privileging distributed interaction through local actions that have a global effect. By shifting the focus from the coordination infrastructure back to the components, and yet retaining a coordination perspective that separates behavior from communication, it is our contention that the notion of GVDS is the right match for the fluid environment defined by current trends in distributed computing.

It is our hope that the ideas presented in this paper will stimulate interest among researchers and cooperation in finding answers to the issues above, thus leading to the definition of a new generation of coordination models and middleware targeting highly dynamic environments.

Acknowledgments   This paper is based upon work supported in part by the National Science Foundation (NSF) under grants NSF CCR-9970939 and EIA-0080124, and by the Italian government under the MIUR project SAHARA.

REFERENCES

1. http://www.napster.com.

2. http://www.gnutella.org.

3. http://freenet.sourceforge.net.

4. G. Berry and G. Boudol. The chemical abstract machine. *Theoretical Computer Science*, 96:217–248, 1992.

5. U. Cetintemel, P.J. Keleher, and M. J. Franklin. Support for speculative update propagation and mobility in deno. In *Proceedings of the IEEE Intl. Conf. on Distributed Computing Systems (ICDCS)*, Mesa, AZ, USA, 2001.

6. D. Chen, C. Tang, X. Chen, S. Dwarkadas, and M.L. Scott. Beyond S-DSM: Shared state for distributed system. In *Proceedings of the International Conference on Parallel Processing (ICPP)*, August 2002.

7. X. Chen and A. Murphy. Enabling disconnected transitive communicaiton in mobile ad hoc networks. In *Workshop on Principles of Mobile Computing, colocated with PODC'01*, pages 21–27, Newport, RI (USA), August 2001.

8. G. Cugola and G.P. Picco. PEERWARE: Core middleware support for peer-to-peer and mobile systems. Technical report submitted for publication. Available at www.elet.polimi.it/~picco.

9. D. Gelernter. Generative Communication in Linda. *ACM Computing Surveys*, 7(1):80–112, Jan. 1985.

10. J.J. Kistler and M. Satyanarayanan. Disconnected Operation in the Coda File System. *ACM Trans. on Computer Systems*, 10(1):3–25, 1992.

11. C. Mascolo, L. Capra, S. Zachariadis, and W. Emmerich. XMIDDLE: A data-sharing middleware for mobile computing. *Int. Journal on Wireless Personal Communications*, April 2002.

12. A.L. Murphy, G.P. Picco, and G.-C. Roman. LIME: A Middleware for Physical and Logical Mobility. In F. Golshani, P. Dasgupta, and W. Zhao, editors, *Proceedings of the 21st International Conference on Distributed Computing Systems (ICDCS-21)*, pages 524–533, May 2001.

13. G.A. Papadopoulos and F. Arbab. Coordination models and languages. In *761*, page 55. Centrum voor Wiskunde en Informatica (CWI), ISSN 1386-369X, 31 1998.

14. G.P. Picco, A.L. Murphy, and G.-C. Roman. LIME: Linda Meets Mobility. In D. Garlan, editor, *Proceedings of the 21st International Conference on Software Engineering (ICSE'99)*, pages 368–377, Los Angeles, CA, USA, May 1999. ACM Press.

15. G.P. Picco, A.L. Murphy, and G.-C. Roman. Developing Mobile Computing Applications with LIME. In M. Jazayeri and A. Wolf, editors, *Proceedings of the 22th International Conference on Software Engineering (ICSE 2000)*, pages 766–769, Limerick (Ireland), June 2000. Formal research demo.

16. D.S. Rosenblum and A.L. Wolf. A Design Framework for Internet-Scale Event Observation and Notification. In *Proc. of the 6th European Software Engineering Conf. held jointly with the 5th Symp. on the Foundations of Software Engineering (ESEC/FSE97)*, LNCS 1301, Zurich (Switzerland), September 1997. Springer.

17. J. Waldo, G. Wyant, A. Wollrath, and S. Kendall. A Note on Distributed Computing. In J. Vitek and C. Tschudin, editors, *Mobile Object Systems: Towards the Programmable Internet*, volume 1222 of *LNCS*, pages 49–66. Springer, April 1997.

18. H. Yu and A. Vahdat. The costs and limits of availability for replicated services. In *Proceedings of the Eighteenth ACM Symposium on Operating Systems Principles (SOSP)*, October 2001.

Author(s) affiliation:

- **Gian Pietro Picco**

  Dipartimento di Elettronica e Informazione
  Politecnico di Milano
  20133 Milano, Italy
  Email: picco@elet.polimi.it

- **Amy L. Murphy**

  Department of Computer Science
  University of Rochester
  Rochester, NY, 14627, USA
  Email: murphy@cs.rochester.edu

- **Gruia-Catalin Roman**

  Department of Computer Science
  Washington University
  St. Louis, MO 63130, USA
  Email: roman@cs.wustl.edu

3

Models of Coordination and Web-based Systems

Robert Tolksdorf

Abstract

We review several coordination models from various disciplines. We use a set of characteristics of coordination models to compare the reviewed ones. As an example of technology for Web-based systems that is based on one of these models, we describe an extension to the Linda model of coordination. It allows XML documents to be stored in a coordination space from where they can be retrieved based on multiple matching relations amongst XML documents, including those given by XML query-languages. This paper is based on the material presented in [TG01a] and [Tol00b].

3.1 LOOKING AT MODELS OF COORDINATION

Todays software is structured into modules, objects, components, agents etc. These entities capture rather small conceptual abstractions and functionalities. While this is advantageous for the design and implementation of software, networked environments add additional benefits when running programs composed from those entities.

It is the interaction of computing units that makes their composition useful. Enabling for a useful interaction is coordinated activity. And in order to provide technology that supports the interaction and its design, models of coordination are necessary. These models have to have certain qualities such as being complete with respect to

31

interaction forms and open to new patterns of interactions. They have to be easy and safe to use to facilitate efficient software engineering. They must be scalable and efficient to implement to cope with the number of units to coordinate. And finally, the models have to be aware of the characteristics of future environments, eg. be robust to failures and dynamics.

Coordination models are enabling for the execution and design of coordinated applications. If it is possible to find commonalities amongst coordination models, then coordination patterns used in the various disciplines could be made transferable. We take a first step towards that goal and look at a variety of existing coordination models from different sources. We evaluate them with respect to a set of qualities.

3.2 MODELS OF COORDINATION

Various disciplines have developed their own models on how entities interact, and do in part also provide technologies that take advantage of these models. Examples are:

- Daily life naturally suggests that independent entities cooperate. Thus, *everyone* has some – diffuse – understanding of what coordination is and should at least be able to tell uncoordinated behaviors from coordinated ones.

- In *computer science*, parallel and distributed programming has been one of the first fields in which coordination models were developed [CC91].

- *Distributed AI* is concerned with the design of coordination in groups of agents and uses a variety of models.

- *Organization theorists* predict the behavior and performance of organizations. As in these the interaction of actors is a key factor, coordination models

- *Economics* look at how interactions in markets takes place and model them.

- *Sociologists* and *Psychologists* explain the behavior of groups and individuals [GM95].

- *Biologists* study phenomena in natural agent systems such as ant colonies or swarms and try to discover the embodied coordination mechanisms [BDT99].

This incomplete list indicates that there is a huge variety of coordination models. We examine some of them in the following.

3.2.1 A Naive Model

A common understanding of coordination is that active entities coordinate to achieve a common goal. This understanding assumes a common goal shared by all entities. It can be explicitly stated and represented, but can also be implicit. The entities can be assumed to be willing to cooperate, that is to follow the goal. They can do so explicitly by cooperating, but this is not required. The naive model provides no clear focus on coordination. Coordination is not encapsulated outside the agents. Coordination mechanisms and policies are therefore not interchangeable.

3.2.2 Mintzberg Model

[Min79] is a seminal work on structures of organizations, coordination mechanisms used therein and dominant classes of configurations of organizations.

Mintzberg develops a theory on the structure of organizations by postulating five basics parts which can be found in any organization. As depicted in figure 3.1, at the basis of the organization is the *operating core* where the actual work is performed. At the (hierarchical) top of organizations is the *strategic apex* – managers who have the overall responsibility for the organization and that take strategic decisions and guide the direction of the organization. The *middle line* is a chain of managers that implement the decisions by supervising subordinates and reporting to their supervisors. The *technostructure* serves to analyse and organize the work done. *Support staff* includes all the indirect support of work, eg. by running a plant cafeteria.

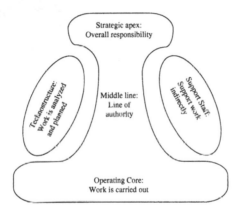

Fig. 3.1   The structure or organizations [Min79]

Coordination in organizations is explained by five mechanisms (table 3.1):

1. *Mutual adjustment* builds on informal communication amongst peers. There is no outside control on decisions and peer coordinate their work themselves.

2. With *Direct supervision*, a supervisor coordinates the work of its subordinates by giving instructions.

3. *Standardization of work* ensures coordination by specifying the work to be done so that no decisions have to be taken later.

4. *Standardization of outputs* refers to specifying the result of work, thus these can be used by others without additional coordination.

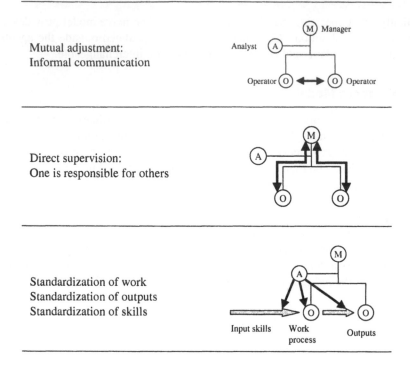

Mutual adjustment:
Informal communication

Direct supervision:
One is responsible for others

Standardization of work
Standardization of outputs
Standardization of skills

Table 3.1   The five coordination mechanisms in organizations [Min79]

5. *Standardization of skills* specifies the training necessary for a work. No further coordination is needed, as peers know what to expect from each other.

Mintzberg selects five typical configurations of coordination mechanism and preeminent parts in the organization as in table 3.2.

Name	Coordination mechanism	Key part
Simple structure	Direct supervision	Strategic apex
Machine Bureaucracy	Standardization of work processes	Technostructure
Professional bureaucracy	Standardization of skills	Operating core
Divisionalized form	Standardization of outputs	Middle line
Adhocracy	Mutual adjustment	Support staff

Table 3.2   The five structural configurations of organizations [Min79]

34

The Mintzberg model assumes a role model for actors in an organization. The choice of coordination mechanisms is induced by the choice of the organizations structure and thus not easily exchangeable. Actors are very aware of the coordination mechanism they have to use. Mintzberg discusses patterns of transitions amongst the different configurations, but these tend to be rather slow.

3.2.3 The Coordination Theory Model

[MC94] introduces the term *coordination theory* to "refer to theories about how coordination can occur in diverse kinds of systems." It draws on different disciplines such as computer science, organization theory, management science and others.

The model defines coordination as *the management of dependencies amongst activities*. In order to make an interdisciplinary use of coordination mechanisms found in various kinds of systems, the processes involved in the management of dependencies have to be studied.

In order to do so, the kinds of dependencies have to be analyzed and the respective processes be studied [Cro91, Del96].

The model assumes that the dependencies are external to activities studied. It is not the activities that are managed, but relations amongst them. The model abstracts from the entities that perform activities and their goals.

3.2.4 Formal Models

In this subsection, we briefly look at formal models of coordination. We follow the overview in [Oss99] and refer to this source for a closer description.

In centralized formal models, there is a known global set of entities to be coordinated. The state of each with respect to coordination activities is modeled by a decision variable. Thus, the system to be coordinated is represented by a set of decision variables $V = \{v_1, \ldots, v_n\}$ (with values from a set of domains $D = \{D_1, \ldots, D_n\}$). Any coordination process leads to an instantiation x of decision variables from decision space X.

In quantitative formal models, a global utility function $U : X \to \Re$ is associated with each of these instantiations that models how good the system is coordinated. The model can be analyzed to find an optimum $y \in X$ such that $\forall x \in X : U(x) \leq U(y)$, meaning that there is no instantiation that provides better coordination.

For a qualitative mode, a set D of constraints $\{C_1, \ldots, C_m\}$ is used. The notion of a consistent instantiation x of decision variables from decision space is defined by X as $y \vDash C_1 \wedge \ldots \wedge C_m$.

Game theory provides a model of decentralized cooperation. The set of entities to be coordinated is modeled as a game which consists of a set I of n players. There is a space S of joint strategies $S = S_1 \times \ldots \times S_n$. It collects the individual strategy $S_i = \sigma_{i_1}, \ldots, \sigma_{i_m}$ that each player has.

In contrast to the global utility function of the quantitative model above, a set P of payoff functions is defined for each player individually by $P_i : S \to \Re$. Two kinds of games are distinguished. In zero-sum games the payoff of one player is "financed" by lower payoffs of the others: $\forall \sigma \in S : \sum_{i=1}^{n} P_i(\sigma) = 0$. In non-constant sum games, this restriction does not exist: $\exists \sigma, \sigma' \in S : \sum_{i=1}^{n} P_i(\sigma) \neq \sum_{i=1}^{n} P_i(\sigma')$.

The situation can be analyzed in two ways. In non-cooperative analysis, the players try to get the best payoff they can individually. The set of strategies is said to be in a Nash equilibrium, if deviation from it by one player will not increase that players payoff: $\forall i \in I : \forall \sigma_i \in S : P_i(\sigma_1^*, \ldots, \sigma_i, \ldots, \sigma_n^*) \leq P_i(\sigma_1^*, \ldots, \sigma_i^*, \ldots, \sigma_n^*)$. In a cooperative analysis, the players coordinate strategies and join payoffs. The situation is said to be Pareto-optimal if no one can achieve a higher payoff without lowering that of some other player.

[RZ94] use game theory to analyse negotiations with societies of agents. These negotiations are categorized into three domains, task-, state- and worth-oriented, which are characterized by different strategies of agents and different entities modelled for coordination.

The formal models share some assumptions. They make a fundamental assumption about the environment, namely that payoff and utility can be defined exact and static. And they assume that agents behave rational to maximize utility or payoff.

3.2.5 Coordination Mechanisms in DAI/MAS

[Jen96] asserts that the key to understanding coordination processes is to look at the internal structures of agents. There, commitments – pledges of agents about actions and beliefs in the future or the past – and conventions – general policies on reconsiderations of commitments – are determent for coordination mechanisms. In addition, social conventions give policies on interactions in a community of agents and local reasoning is necessary to use that knowledge.

In Computational Organization Theory, roles are defined that constrain behavior of agents. In Multi-agent Planning, commitments are based on plans that agents develop. In a multi agent society setting, negotiation is the coordination mechanism by which agents take joint decisions after following some negotiation protocol.

The models assume that the conventions governing the coordination processes are external to the agents. They assume commitments a priori to events that take place, and thus assume knowledge about future events.

3.2.6 Uncoupled Coordination

In parallel computing, questions on how to organize the execution of multiple concurrent threads in a computation has led to several coordination models. The model of the language Linda [GC92] takes the view that coordination has to be performed explicitly by the parallel processes and that it is worthwhile to use a separate language for that. Such a coordination language focuses merely on the expression of coordination and defines a respective coordination medium.

The language Linda embodies a model of uncoupled coordination where a set of agents together form an ensemble in which they coordinate their interaction indirectly by using a shared dataspace, called the *tuplespace*.

The coordination language is defined in terms of operations that access and modify the tuplespace. See section 3.5 for their description.

The model assumes explicit coordination amongst agents that have to use the coordination language in an appropriate way. The pattern of coordination is scattered over the use of the coordination operations with the agents.

The model provides an abstraction from the location of agents in space and time. Agents remain anonymous to one other and do not necessarily have to exist at the same time. Also, it abstracts from the computational model and programming language used by the agents.

3.2.7 Workflow

Workflow Management Systems (WfMS) coordinate human work and its support by applications. In recent years, there has been an urge to define a common denominator of such workflow modeling languages to enhance interoperability amongst systems of different vendors. The Workflow Management Coalition (WfMC) is the industry consortium of the leading WfMS vendors and has published a reference model. Part of it is a process definition language [Wor98a] that represents a minimal language to express workflow models.

Here, a workflow is modeled as a graph of activities as nodes and transitions between them. The transitions represent dependencies amongst activities and can be augmented with additional constrains, such as start and end-times. The topology of the graph includes coordination constraints on activities.

So called AND-JOIN- and AND-SPLIT-nodes synchronize activities or span new parallel activities. An XOR-JOIN makes the execution of an activity dependent on the termination of one out of several other ones. XOR-SPLIT selects one new thread of activities to start. The flow of activities can be further specified by introducing loops and sub-workflows. As shown in figure 3.2, activities are performed by participants in the workflow and might involve data and applications. The participants are constituent for the organization in which the workflow takes place.

The approach taken by most WfMS assumes that all activities and dependencies amongst them are known in advance. Also, reliable execution of activities is assumed – at least the reference model lacks the notion of exceptions.

3.3 COMPARING COORDINATION MODELS

The models reviewed can be explored by a set of characteristics, as follows.

- To what degree the model provides a *clear distinction* amongst interactors, non-interactors and management of relations.

- How *orthogonal* the coordination model is with computational models used.

37

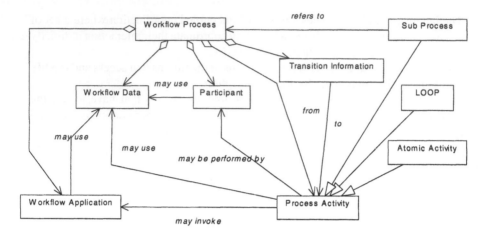

Fig. 3.2 The reference model of the WfMC

- The degree of *coupling* between interactors and to an external goal.

- The degree of *autonomy* of interactors, inverse to the degree of centralization.

- Whether the management of relation is *external* to interactors or not.

- How much *awareness* to management of relation is required from interactors.

- The degree of *stability of interactors* assumed by the model on the interactors can be used as programming model.

- Existence of qualitative or quantitative *measures* on management of relations.

Further dimensions seem useful but have not been considered here. Examples are realtime aspects or how a model views the environment of interactors, ie. the situation of the agents. With respect to the dimensions listed, tables 3.3 and 3.4 show a comparison of the models reviewed.

The overview thus shows that each model can described by dominant characteristics along our dimensions, which are shown boldfaced in the table. Also, it shows that the models reviewed here cover very well the scale of those dimensions.

We now look closer at one specific application of the Linda-model for Web-based systems, XMLSpaces.

3.4 WEB-BASED SYSTEMS

While the Web has become *the* universal information system worldwide in the first decade of its existence, the progress towards cooperative information systems utilizing

Model	Distinction	Orthogonal	Coupling	Autonomy	External	Awareness	Interactors stability
Naive	No	Yes	High	**Low:** Shared goal	No: Entities coordinate	Yes	Midterm: Until goal reached
Mintzberg	Yes: Inputs and outputs in addition to actors	Yes	Dep. on mechanism	Low	No	Yes	Midterm
Coord. Theory	Yes	Yes	**Dep. on mechanism**	Mid	Yes	Yes	Midterm
Formal	No	**No:** Evaluation of functions in model	High	Low	No	Yes	Longterm
DAI/MAS	Yes	Yes	Mid	High	Yes	**Yes**	Midterm
Linda	Yes	Yes	**Low**	Mid	Yes	Mid	**Shortterm**
WfMC	Yes	Yes	High	Low	Yes	Yes	**Longterm**

Table 3.3 Comparing the coordination models

Model	Relations stability	Reliability	Scalability	Programming	Measures
Naive	Midterm: Until goal reached	Yes	No	No	No
Mintzberg	**Longterm**	No	Yes	Unclear	No
Coord. Theory	Midterm: Until dissolved	Yes	No	Unclear	No
Formal	Longterm	Yes	No	Yes	**Yes**
DAI/MAS	Midterm: Horizon of reasoning	No	**No**	Yes: Agent systems	No
Linda	**Shortterm**	Yes (see [TR00])	Unclear (see [MTW00])	Yes	No
WfMC	**Longterm**	Yes	No	Yes	Yes

Table 3.4 Comparing the coordination models (cont.)

the Web for universal access is rather slow. Although there are several technologies like Java or CORBA available, none of these has reached universal acceptance.

Recently, the Web-Standard XML (*Extensible Markup Language*) [Wor98c] has become *the* format to exchange data markup following application specific syntaxis. It will be the dominating interchange format for data over networks for the next years. XML data is semi-structured and typed by an external or internal document type or by a minimal grammar inferred from the given document. A DTD (*Document Type Definition*) defines a context-free grammar to which an XML document must adhere. Tags define structures within a document that encapsulate further data. With attributes, certain meta information about the data encapsulated can be expressed. While XML enables collaboration in distributed and open systems by providing common data formats, it is still unclear how components coordinate their work.

The concept of *XMLSpaces* presented in this paper marries the common communication format XML with the coordination language Linda [TG01b]. It aims at providing coordination in Web-based cooperative information systems. XMLSpaces offers a simple yet flexible approach to coordinate components in that context and extends the original Linda-notion with a more flexible matching concept.

40

3.5 LINDA-LIKE COORDINATION

Linda-like languages are based on data-centric coordination models. They introduce the notion of a shared dataspace that decouples partners in communication and collaboration both in space and time [CG89].

The coordination media in Linda is the *tuplespace* which is a multiset of *tuples*. These are in turn ordered lists of unnamed fields typed by a set of primitive types. An example is ⟨10,"Hello"⟩ which consists of an integer and a string.

The tuplespace provides operations that uncouple the coordinated entities in time and space by indirect, anonymous, undirected and asynchronous communication and synchronization. The producer of data can emit a tuple to the tuplespace with the operation *out(⟨10,"Hello"⟩)*. The consumer of that data does not even have to exist at the time it is stored. The producer can terminate before the data is consumed.

To consume some data, a process has to describe what kind of tuple shall be retrieved. This description is called a *template*, which is similar to tuples with the exception, that fields also can contain bottom-elements for each type, eg. ⟨10,?string⟩. These placeholders are called *formals* in contrast to *actuals* which are fields with a value. Given a template, the tuplespace is searched for a *matching* tuple. A *matching relation* on templates and tuples guides that selection.

Retrieving a matching tuple is done by *in(⟨10,?string⟩)* which returns the match and removes it from the space. The primitive *rd(⟨10,?string⟩)* also returns a match but leaves the tuple in the tuplespace. Both primitives *block* until a matching tuple is found, thereby synchronizing the consumer with the production of data.

The Linda matching relation requires the same length of tuples and templates and identical types of the respective fields. For formals in the template, the actual in the tuple has to be of same type, while actuals require the same value in the tuple.

Tuples as in Linda can be considered "primitive data" – there are no higher order values such as nested tuples, no mechanisms to express the intention of typing fields such as names etc. For coordination in Web-based systems, a richer form of data is needed. It has to be able to capture application specific higher data-structures easily without the need to encode them into primitive fields. The format has to be open so that new types of data can be specified. And it has to be standardized in some way, so that data-items can be exchanged between entities that have different design-origins.

The *Extensible Markup Language* XML [Wor98c] has recently been defined as a basis for application specific markup for networked documents. It seems to meet all the outlined requirements as a data-representation format to be used in a Linda-like system for open distributed systems. *XMLSpaces* is our system that uses XML documents in addition to ordinary tuple fields to coordinate entities with the Linda-primitives.

3.6 XMLSPACES

XMLSpaces extends the Linda model in several major aspects:

1. XML documents serve as fields within the coordination space. Ordinary tuples are supported, while XML documents can be represented as one-fielded tuples.

2. A multitude of relations amongst XML documents can be used for matching. The system is open for extension with further relations.

3. XMLSpaces is distributed so that multiple dataspace servers at different locations form one logic dataspace. A clearly separated distribution policy can easily be tailored to different network restrictions.

4. Distributed events are supported so that clients can be notified when a tuple is added or removed somewhere in the dataspace.

We describe these extensions in the following. In XMLSpaces, actual tuple fields can contain an XML document, formal fields can contain some XML document description, such as a query in an XML query language. The matching relation is extended on the field-field level with relations on XML documents and expressions from XML query languages. All Linda operations, and the matching rule for other field-types and tuples are unchanged.

The matching rule to use for XML fields is not statically defined, instead, XMLSpaces supports multiple matching relations on XML documents. The current implementation of XMLSpaces builds on a standard implementation of Linda, namely TSpaces [WMLF98]. It already provides the necessary storage management and the basic implementations for the Linda primitives.

In TSpaces, tuple fields are instances of the class *Field*. It provides a method called *matches(Field f)* that implements the matching-relation amongst fields and returns *true* if it holds. The method is called by the matching method of class *SuperTuple*, which tests for equal length of tuples and templates. Actuals and formals are not modeled as distinguished classes but typed according to their use in matching.

XMLSpaces introduces the class *XMLDocField* as a subclass of *Field*. The contents of the field is *typed* as an actual or a formal by fulfilling a Java-interface. If it implements the interface *org.w3c.dom.Document*, it is an actual field containing an XML document. If it implements the interface *XMLMatchable*, it is a formal. Otherwise it is an invalid contents for an *XMLDocField*.

The method *matches* of an *XMLDocField* object tests the polarity of fields for matching. It returns false, when both objects are typed as formals, or when an actual is to be matched against a formal. If both the *XMLDocField*-object and the parameter to *matches* are actuals, a test for equality is performed. Otherwise – if a formal is to be matched against an XML document – the method *xmlMatch* of the formal is used to test a matching relation. Figure 3.3 shows the resulting class hierarchy.

3.6.1 Multiple Matching Relations

The purpose of the interface *XMLMatchable* is to allow for a variety of matching relations amongst XML documents. The template used for *in* and *rd* then, is not relative to the language definition as with Linda, but relative to a relation on XML documents

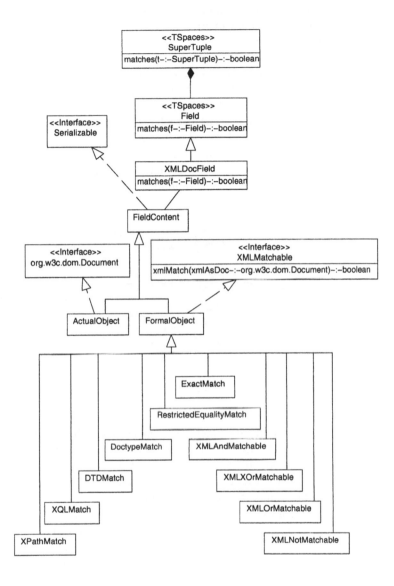

Fig. 3.3 The class hierarchy for XML documents in tuple fields in UML notation

and XML templates that is contained within the template as the implementation of *xmlMatch* in *XMLMatchable*.

The use of multiple matching relations can be an application requirement. We find such a requirement in the Workspaces architecture [Tol00a]. Workspaces is a

Web-based workflow system which combines concepts from the application domain workflow management with standard Internet technology, namely XML and XSL, and with coordination technology.

Steps are the basic kind of activity in Workspaces and represent a unit of work on some application specific XML document. Each step is represented as an XML document that can be distributed individually for interpretation by the XSL-based Workspaces engine. As a result, distributed workflows can be coordinated via the Web. A complete Workflow is described as a graph in an XML-document. It is split into a set of individual steps in an XSL-based compilation step.

Figure 3.4 shows an example of such a workflow graph – in this case describing the review process for papers submitted to a conference. The application specific documents being manipulated are the papers submitted and reviews forms to be filled out by members of a program committee.

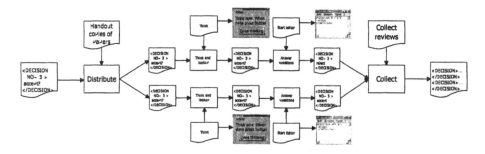

Fig. 3.4 A WorkSpaces workflow for reviewing conference submissions

Each step is described in another document, the step document. It contains an XSL script interpreted by the WorkSpaces engine to automatically transform documents, eg. the "Collect reviews" step, to call external applications as with the "Answer questions" step, or to wait for events external to the system as seen for the step "Think and lookup". XSL scripts are valid XML documents.

The Workspaces engine utilizes XMLSpaces as shown in figure 3.5. First, some work description is retrieved with an *in* by requesting an XML document that matches the step DTD. Then, the necessary application specific XML document is requested by referring to some identifier in an attribut of a tag of the document. Then, the actual work is performed as described by the step document. Finally, the changed document is put back to the XMLSpaces with an *out* operation.

During execution of a workflow, one might try to retrieve "*something* to do", which means a document that follows the DTD used for the description of steps. If, however, a specific task is to be done on a specific application document, one wants that *one* XML document that might be described by an identifier in an attribute. This requirement induces the need for support of multiple relations used in matching.

The Workspaces engines benefit from the application of this kind of coordination technology. They are completely uncoupled and can be distributed and mobile. The number of engines participating in the system can be dynamic as new engines can

44

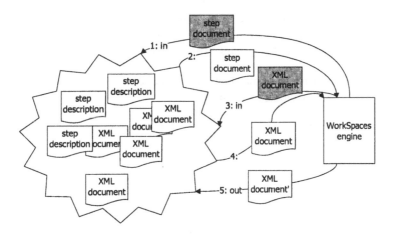

Fig. 3.5 Usage of XMLSpaces in Workspaces

join and leave at whatever time and location they want. The engine that will process future steps does not necessarily have to run when the workflow starts to execute. These attractive advantages of our architecture are due to the use of coordination technology and indicate its usefulness. The decoupled coordination style, indirected by a coordination medium that masks any issues of distribution and synchronization, gives huge technical freedom for a distributed and open implementation.

XMLMatchable is the basis of the extensibility of XMLSpaces with new matching relations. To realize it, a new class has to be provided that implements this interface and tests for the new relation in the *xmlMatch* method. Figure 3.6 shows an implementation of one matching routine. which uses the XQL-engine from GMD-IPSI. It shows the ease of integration of further matching routines in XMLSpaces.

```
package matchingrelation;
import xmlspaces.XMLMatchable;
import java.io.*;
import org.w3c.dom.Document;
import de.gmd.ipsi.xql.*;
public class XQLMatch implements XMLMatchable {
  String query;
  public XQLMatch(String xqlQuery) {
    query = xqlQuery;
  }
  public boolean xmlMatch(Document xmlAsDoc) {
    return XQL.match(query, xmlAsDoc); // forward to XQL engine
  }
}
```

Fig. 3.6 The implementation of XQL-matching

45

While the XML standard defines one relation, namely "validates" from an XML document to a DTD, there is a variety of possible other relations amongst XML documents and other forms of templates. These include:

- An XML document can be matched to another one based on equality of contents, or on equality of attributes in elements.

- An XML document can be matched to another one which validates against the same grammar, ie. DTD.

- An XML document can be matched to another one which validates against the same minimal grammar with or without renaming of elements and attributes.

- An XML document can be matched to a query expression following the syntax and semantics of those, for example XML-QL, XQL, or XPath/Pointer.

Relation	Meaning
Exact equality	Exact textual equality
Restricted equality	Textual equality ignoring comments, PI's, etc.
DTD	Valid towards a DTD
DOCTYPE	Uses specific Doctype name
XPath	Fulfills an XPath expression
XQL	Fulfills an XQL expression
AND	Fulfills two matching relations
NOT	Does not fulfill matching relation
OR	Fulfills one or two matching relations
XOR	Fulfills one matching relation

Table 3.5 Matching relations in XMLSpaces

Currently, several relations are implemented in XMLSpaces as shown in table 3.5. The relations fall into different categories:

- The *equality* relations use several views on what equality of XML documents actually means. Eg., whether comments are included in a check or not.

- The *DTD* relations take the relation of a document to a DTD or a doctype name as constituent for matching.

- The *query language* relations build on several existing XML oriented query languages. A query describes a set of XML documents to which the query matches. The query match then is taken as the matching relation in the sense of XMLSpaces. Note that the query languages are of high expressibility, for example, matching for all documents that contain a specific value in some

46

attribute can be formulates as an XPath or XQL expression. While the equality and DTD relations consider a document as a whole, the query relations try to find a match in one part of a document.

- The *connector* relations allow it to build boolean expressions on matching relations whose result gives the final matching relation.

3.6.2 Distributed XMLSpaces

XMLSpaces supports the integration of XMLSpaces servers at different places into a single logic dataspace. Distribution of a Linda-like system can be implemented using different distribution schemata which have different efficiency characteristics:

- *Centralized*: One server holds the complete dataspace.

- *Distributed*: All servers hold distinct subsets of the complete dataspace.

- *Full replication*: All servers hold consistent copies of the complete dataspace.

- *Partial replication*: Subsets of servers hold consistent copies of subsets of the dataspace [Fra91].

- *Hashing*: Matching tuples and templates are stored at the same server selected by some hashing function [Bjo92].

XMLSpaces does not prescribe one specific approach but encapsulates the distribution strategy in a *distributor object*. It handles the registration of a server in the distributed space and offers distributed versions of the coordination primitives.

The implementation of the distributor object implements a distribution strategy by the respective versions of these methods. XMLSpaces servers are configured with a distribution strategy at startup. At that time, a respective *Distributor* object is created for the XMLSpaces server, whose *register* method makes the server known to remote ones. Exact details of this process are specific to the chosen distribution strategy.

XMLSpaces is *open* in the sense that, with a suited distribution strategy, servers can join and leave at any time. As the distributor object has to know about registered servers, its interface includes methods to register and deregister remote servers.

Currently, XMLSpaces includes implementations of the centralized and partial replication strategy. The system can easily be extended by other distributor objects that implement other strategies. The choice of the distribution policy is configured at startup in a configuration file.

The nodes in the distributed XMLSpaces each store a subset of the complete contents of the data store. The nodes organized in so called *out-sets* contain identical replicas of a subset. An *out*-operation transmits the argument data to all members of the out-set for storage. Every node is at the same time a member of a so called *in-set*. The nodes within one in-set store different subsets of the complete dataspace and the union of their contents represents the whole dataspace. Thus, any *in-* or *rd*-operation

47

asks all nodes in the in-set for a match. In contrast to [Tol98], XMLSpaces does not use a software-bus but members of a set communicate point-to-point.

The structure formed by the sets has to be rectangular – a condition that cannot be upheld in the case of open systems with a varying number of participating nodes. Therefore, the structure has to be retained by *simulating* nodes if necessary. One physical node then is part of two out- or in-sets. The reconfiguration of the system in the case of joining and leaving nodes is part of the protocol for joining and leaving nodes.

In XMLSpaces, there is exactly one special node, the *receptionist*, that decides one-at-a-time how new nodes join the structure. It can but does not have to be colocated with a tuplespace server. The receptionist knows about the complete structure of nodes in the system. When a new node joins, it asks the receptionist for its place in the network. Based on heuristics on the fraction of simulated nodes or considerations about the communication efficiency in the in- and out-sets, the receptionist decides on a place in the grid-structure. It informs the new node about the nodes in the target in- and out-sets and on the address of a simulated node that it shall replace.

It turns out, that the strategies distributed and full replication as mentioned above are special cases of partial replication: The distributed strategy uses only one in-set while full replication uses only one out-set. They can be implemented by changing the decision of receptionist on the placement of new nodes in that respective one set. XMLSpaces offers respective subclasses of the *Receptionist* class. There is no need to introduce additional subclasses of *Distributor*.

3.6.3 Distributed Events

TSpaces supports *events* that can be raised when a tuple is entered or removed from the dataspace. XMLSpaces extends this mechanism to support *distributed events* where clients can register for an event occuring somewhere in the distributed dataspace. As working with distributed events depends on the distribution strategy used, it is implemented in the *Distributor* object.

Distributed events require mechanisms to register for events, to unregister, notifying about events and handling registered events when integrating new nodes into the grid-structure. In the case of partial replication, registering and deregistering for events has to be done on all nodes of the in-set. Events are delivered locally to clients or forwarded to servers in the in-set, that inform their registered clients.

When a new node joins the system by replacing a simulated node, it has to copy all event registrations along with the state. If it forms a new in-set, there are no registrations that have to be considered. Finally, if it forms a new out-set, it has to synchronize with all event registrations on all of its in-sets.

When leaving a system, all locally and remotely registered events have to be deregistered. If the node was the last in its out-set, the registrations can simply be deleted, as no more events will happen. If the leaving node will be simulated afterwards, all registered events have to be transferred to the simulating node.

As with the distribution strategy, it turns out, that distributed and full replication are special cases of partial replication also with respect to events and thus, the distributor implementation can remain unchanged.

3.6.4 Implementation

XMLSpaces extends the original Linda conception with XML documents and distribution. It does not change the set of primitives supported nor affect the implemented internal organization of the dataspace. Thus, we have chosen to build on an existing Linda-implementation, namely TSpaces [WMLF98].

In order to support distribution, the original TSpaces implementation had to be extended at some places. TSpaces allowed for a rapid implementation of XMLSpaces focusing on the extensions. However, it could well be exchanged by some other extensible Linda-kernel.

The standard document object model DOM [Wor98b], level 1, serves as the internal representation of XML documents in actual fields. This leads to a great flexibility to extend XMLSpaces with further matching relations using a standard API. It has shown that the integration of such an engine into XMLSpaces is extremely simple when written in Java and utilizing DOM. If not, some wrapper-object has to be specified in addition. XMLSpaces itself is completely generic towards how the *xmlMatch*-method is implemented and what its semantics are.

As seen in table 3.5, the huge amount of XML related software provided engines that could directly evaluate the relations on XML documents we are interested in.

3.6.5 Related Work

There are some projects documented on extending Linda-like systems with XML documents. However, XMLSpaces seems to be unique in its support for multiple matching relations and its extensibility.

MARS-X [CLZ00] implements an extended JavaSpaces [FHA99] interface. Tuples are represented as Java-objects where instance variables correspond to tuple fields. Such an tuple-object is represented as an element within an XML document. Its representation has to follow a tuple-specific DTD. MARS-X closely relates tuples and Java objects and does not look at arbitrary relations amongst XML documents.

XSet [ZJ00] is an XML database which also incorporates a special matching relation amongst XML documents. Here, queries are XML documents themselves and match any other XML document whose tag structure is a strict superset of that of the query. It should be simple to extend XMLSpaces with this engine.

The note in [Mof99] describes a preversion for an XML-Spaces. However, it provides merely an XML based encoding of tuples and Linda-operations with no significant extension. Apparently, the proposed project was not finished up to now.

TSpaces has some XML support built in [WMLF98]. Tuple fields can contain XML documents which are DOM-objects generated from strings. The *scan*-operation takes an XQL query and returns all tuples that contain a field with an XML document in which one or more nodes match the query. This ignores the field structure and does not follow the original Linda definition of the matching relation.

3.7 CONCLUSION AND OUTLOOK

We have seen that the variety of coordination models is large and draws on various disciplines. With respect to to a set of characteristics of models, we found that the models are well-distinguishable along several dimensions. We found that each model has a dominant characteristic. We also found that the set of models covers substantial parts on the scales of the dimensions considered.

XMLSpaces is a distributed coordination platform that extends the Linda coordination language with the ability to carry XML documents in tuple fields. It is able to support multiple matching relations on XML documents. Both the set of matching relations and the distribution strategy are extensible.

XMLSpaces satisfies the need for better structured coordination data in the Web context by using XML in an open end extensible manner. It has shown that the Linda concept can be extended easily while retaining the original concepts on coordination and a very small core of the coordination language.

http://www.robert-tolksdorf.de/xmlspaces gives further details about XMLSpaces.

Acknowledgment The IBM Almaden Research Center supported the work on XMLSpaces by granting a license to the TSpaces source code.

REFERENCES

BDT99. Eric Bonabeau, Marco Dorigo, and Guy Theraulaz. *Swarm Intelligence*. Oxford University Press, 1999.

Bjo92. Robert Bjornson. *Linda on Distributed Memory Multiprocessors*. PhD thesis, Yale University Department of Computer Science, 1992. Technical Report 931.

CC91. Roger S. Chin and Samuel T. Chanson. Distributed Object-Based Programming Systems. *ACM Computing Surveys*, 23(1):91–124, March 1991.

CG89. Nicholas Carriero and David Gelernter. Linda in Context. *Communications of the ACM*, 32(4):444–458, 1989.

CLZ00. Giacomo Cabri, Letizia Leonardi, and Franco Zambonelli. XML Dataspaces for Mobile Agent Coordination. In *15th ACM Symposium on Applied Computing*, pages 181–188, 2000.

Cro91. Kevin Ghen Crowston. *Towards a Coordination Cookbook: Recipes for Multi-Agent Action*. PhD thesis, Sloan School of Management, MIT, 1991. CCS TR# 128.

Del96. Chrysantos Nicholas Dellarocas. *A Coordination Perspective on Software Architecture: Towards a Design Handbook for Integrating Software Components.* PhD thesis, Massachusetts Institute of Technology, 1996.

FHA99. Eric Freeman, Susanne Hupfer, and Ken Arnold. *JavaSpaces principles, patterns, and practice.* Addison-Wesley, Reading, MA, USA, 1999.

Fra91. Craig Fraasen. Intermediate Uniformly Distributed Tuple Space on Transputer Meshes. In J.P. Banâtre and D. Le Métayer, editors, *Research Directions in High-Level Parallel Programming Languages*, number 574 in LNCS, pages 157–173. Springer, 1991.

GC92. David Gelernter and Nicholas Carriero. Coordination Languages and their Significance. *Communications of the ACM*, 35(2):97–107, 1992.

GM95. Jonathon Gillette and Marion McCollom, editors. *Groups in Context, A New Perspective on Group Dynamics.* University Press of America, 1995.

Jen96. Nicholas Jennings. Coordination Techniques for Distributed Artificial Intelligence. In G. M. P. O'Hare and N. R. Jennings, editors, *Foundations of Distributed Artificial Intelligence*, pages 187–210. John Wiley & Sons, 1996.

MC94. Thomas W. Malone and Kevin Crowston. The Interdisciplinary Study of Coordination. *ACM Computing Surveys*, 26(1):87–119, March 1994.

Min79. H. Mintzberg. *The Structuring of Organizations: A Synthesis of the Research.* Prentice Hall, Englewood Cliffs, N.J., 1979.

Mof99. David Moffat. XML-Tuples and XML-Spaces, V0.7. http://uncled.oit.unc.edu/XML/XMLSpaces.html, Mar 1999.

MTW00. R. Menezes, R. Tolksdorf, and A.M. Wood. Scalability in LINDA-like Coordination Systems. In Andrea Omicini, Franco Zambonelli, Matthias Klusch, and Robert Tolksdorf, editors, *Coordination of Internet Agents: Models, Technologies, and Applications.* Springer, 2000.

Oss99. Sascha Ossowski. *Co-ordination in artificial agent societies: social structures and its implications for autonomous problem-solving agents*, volume 1535 of *LNCS*. Springer Verlag, 1999.

RZ94. Jeffrey S. Rosenschein and Gilad Zlotkin. *Rules of Encounter: Designing Conventions for Automated Negotiation among Computers.* The MIT Press, Cambridge, Massachusetts, 1994.

TG01a. Robert Tolksdorf and Dirk Glaubitz. Coordinating Web-based Systems with Documents in XMLSpaces. In *Proceedings of the Sixth IFCIS International Conference on Cooperative Information Systems (CoopIS 2001)*, number LNCS 2172, pages 356–370. Springer Verlag, 2001.

TG01b. Robert Tolksdorf and Dirk Glaubitz. XMLSpaces for Coordination in Web-based Systems. In *Proceedings of the Tenth IEEE International Workshops on Enabling Technologies: Infrastructure for Collaborative Enterprises WET ICE 2001*. IEEE Computer Society, Press, 2001.

Tol98. Robert Tolksdorf. Laura - A Service-Based Coordination Language. *Science of Computer Programming, Special issue on Coordination Models, Languages, and Applications*, 1998.

Tol00a. Robert Tolksdorf. Coordination Technology for Workflows on the Web: Workspaces. In *Proceedings of the Fourth International Conference on Coordination Models and Languages COORDINATION 2000*, LNCS. Springer-Verlag, 2000.

Tol00b. Robert Tolksdorf. Models of Coordination. In Andrea Omicini, Robert Tolksdorf, and Franco Zambonelli, editors, *Engineering Societies in the Agent World First International Workshop, ESAW 2000, Berlin, Germany, August 21, 2000*, number LNAI 1972, pages 78–92. Springer Verlag, 2000.

TR00. R. Tolksdorf and A. Rowstron. Evaluating Fault Tolerance Methods for Large-Scale Linda-Like Systems. In Proceedings of the 2000 International Conference on Parallel and Distributed Processing Techniques and Applications (PDPTA'2000), 2000.

WMLF98. P. Wyckoff, S. McLaughry, T. Lehman, and D. Ford. T Spaces. *IBM Systems Journal*, 37(3):454–474, 1998.

Wor98a. Workflow Management Coalition. Interface 1: Process Definition Interchange Process Model, 1998. http://www.wfmc.org.

Wor98b. World Wide Web Consortium. Document Object Model (DOM) Level 1 Specification. W3C Recommendation, 1998. http://www.w3.org/TR/REC-DOM-Level-1.

Wor98c. World Wide Web Consortium. Extensible Markup Language (XML) 1.0. W3C Recommendation, 1998. http://www.w3.org/TR/REC-xml.

ZJ00. Ben Yanbin Zhao and Anthony Joseph. The XSet XML Search Engine and XBench XML Query Benchmark. Technical Report UCB/CSD-00-1112, Computer Science Division (EECS), University of California, Berkeley, 2000. September.

Author(s) affiliation:

- **Robert Tolksdorf**

  Technische Universität Berlin
  Fachbereich 13, Informatik
  D-10587 Berlin
  Email: research@robert-tolksdorf.de

4

Temporal Logic Coordination Models

Chuang Lin
Dan C. Marinescu

Abstract

In recent years a significant body of research has been dedicated to the analysis and verification of workflows without taking into account the temporal dimension of activities involved. Once the temporal aspects of the activities in a process description are taken into account, various types of synchronization anomalies and other undesirable behavior may be detected. In this paper, we use a time Petri nets representation of workflows. We propose temporal logic workflow models and introduce a set of linear inference algorithms for the qualitative and quantitative analysis of temporal constraints of workflows. We apply our models to the analysis of several workflow patterns.

4.1 INTRODUCTION

The term workflow means the coordinated execution of multiple tasks or activities. Production, administrative, collaborative, and ad-hoc workflows require that documents, information or tasks be passed from one participant to another for action, according to a set of procedural rules. Production workflows manage a large number of similar tasks with the explicit goal of optimizing productivity. Administrative

53

workflows define processes, while collaborative workflows focus on teams working towards common goals.

Originally, workflow management was considered a discipline confined to the automation of business processes. Yet the basic ideas and technologies for workflow management are common to areas ranging from science and engineering to entertainment; the collection and analysis of experimental data in a scientific experiment, battlefield management, logistics support for the merger of two companies, health care management, are all examples of complex activities described by workflows.

Nowadays, in addition to the expansion of the scope of workflow management we are witnessing another profound change, this time related to the environment and the underlying technologies for workflow management. In the general case, the actors involved in a workflow are geographically scattered and communicate via the Internet. In such cases reaching consensus amongst various actors involved is considerably more difficult.

Today most business processes are dependent upon the Internet and the workflow management had evolved into a network-centric discipline [17]. Automated process coordination provides the means to improve the quality of service, increase flexibility, allow more choices, support more complex services offered by independent service providers in an information grid. E-commerce and Business-to-Business are probably the most notable examples of Internet-centric applications requiring some form of workflow management.

The workflow management is a complex process including definition, verification monitoring, control, optimization of processes that are often subject to timing constrains. Traditional workflow models support the representation of external events and simultaneous actions and are able to deal with the combination of sequential relationships and concurrency. Workflow models using different flavors of Petri Nets have been used for static and dynamic analysis and for verification of the process description of a workflow, for some time.

Time plays an important role in workflow modeling and analysis. As early as 1980's it was recognized that office automation systems should incorporate the notion of time [9]. More recently, the need to support temporal aspects of workflows such as deadlines, variable calendar windows, time scales, and alerting mechanisms for overdue actions, was reiterated in the context of functional requirements for the next generation of workflow software, [25].

The temporal reasoning and performance evaluation of workflows are research topics yet to be given the importance they deserve. In this paper we analyze workflows subject to timing constraints and introduce timed Petri nets workflow models. We also define linear reasoning algorithms to evaluate temporal properties of basic workflow patterns. We extend the algorithms to complex temporal workflow models composed of basic patterns and show that they is capable to solve real-world problems.

The algorithms introduced in this paper can be used for sound workflow models; they conveniently express any temporal relationships among two tasks, including synchronization and conflict. The time complexity of our algorithms is linear thus they have substantial advantages compared with traditional methods of analysis based upon the reachability graphs [27].

The paper is organized as follows: Section 4.2 provides a general background in the area of workflow modeling with Petri Nets; Section 4.3 introduces temporal logic workflow models; Section 4.4 presents linear reasoning algorithms for basic workflow patterns.

4.2 PETRI NETS MODELS OF WORKFLOWS

In recent years the Workflow Management Coalition made a concerned effort for standardization, specification, and analysis of workflow management systems [26]. In spite of this effort and of the increased interest in workflow management, the field still lacks precise definitions for some of its concepts; little agreement exists upon what a workflow exactly stands for and the specific features a workflow management system must support.

For this treason the efforts to relate workflows with well established models of computations, like Petri nets are very important. Van der Aalst identifies three main reasons for using Petri nets for workflow modeling and specification [1]: the Petri nets possess a formal semantics and an intuitive graphical representation; they can explicitly model states and a clear distinction can be made between the enabling and execution of a task; and the abundance of available and theoretically proven analysis techniques.

For a review of different flavors of Petri Net-based workflow models see [14]. Zisman used Petri nets to model workflow processes for the first time in 1977. In 1993 Ellis and Nutt introduced a flavor of High Level Petri Nets called Information Control Nets, ICN, to model the follow of control and data [11]. The Workflow nets, WF-nets, proposed by van der Aalst [3] are High Level Petri Nets with two special places I and O, indicating the beginning and the end of the modeled process. In general, the WF-net is a marked graph. In the ideal case every transition is on a path, and each path is bounded by a fork and a join transition. A *fork*, is a transition with more than one output places and a *join*, is a transition with more than one input places. The WF-nets are suitable not only for representation and validation, but also for the verification of workflows. To model temporal dependencies between two tasks in a workflow, however, Adam et al. proposed a Temporal Constraint Petri Net, TCPN [6]; each place and each transition of a TCPN are associated with a time interval and a token is associated with a time stamp.

Murata et al. introduced a Fuzzy Timing Petri Net (FTN) model for distributed multimedia synchronization capable to handle fuzzy temporal requirements [19]. FTN introduces four fuzzy time functions called the fuzzy timestamp, fuzzy enabling time, fuzzy occurrence time, and fuzzy delay.

We now discuss Time Petri Nets, TPN, and their applications to workflow modeling. In a TPN model, each transition T has an associated time section, $X_T(t) = [t_l, t_u], (0 \leq t_l \leq t_u)$. Recall that only an enabled transition may fire and the enabling time is controlled by the system. Call s the enabling time; t^* actual firing time; and t_l, and t_u the lower and upper bounds of the interval when the transition fires, once it is enabled:

$$s + t_l \leq t^* \leq s + t_u.$$

This proposition of a TPN model corresponds to a basic workflow task fired by time $X[sx, lx, ux]$. The enabling time s corresponds to sx, and the transition is fired at time t^* during the interval $[lx, ux]$. $sx + t_l$ corresponds to lx; $sx + t_u$ corresponds to ux, see Figure 4.1:

$$t_l = lx - sx, \quad t_u = ux - sx$$

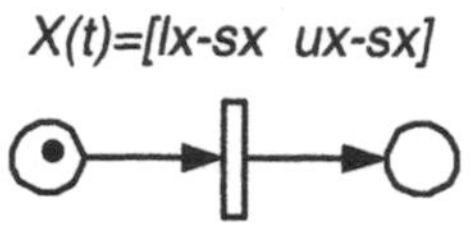

X(t)=[lx-sx ux-sx]

Fig. 4.1 The Time Petri Net representation of a task.

In case of an instantaneous transition $X(t) = [0, 0]$, the transition fires as soon as it is enabled. In a TPN graph, a bar represents an instantaneous transition while a time transition is represented by a rectangle, see Figure 4.1.

4.3 TEMPORAL LOGIC WORKFLOW MODELS

Temporal modeling and analysis of workflows could be based upon several representations and reasoning schemes for temporal information. The Interval Temporal Logic, ITL, is a temporal logic developed by Moszkowski [20] [21]. The qualitative temporal calculus proposed by Allen [7], [8] provides an alternative representation and reasoning scheme for temporal information.

ITL is based upon a flexible notation for both propositional and first-order reasoning about periods of time found in a process descriptions of an workflow. The term *interval* refers to the operational process of an action or event and can be abstracted into a proposition. The phrase *interval temporal logic* abstracts the temporal relations between two time intervals or propositions. Unlike other temporal logic systems, ITL can handle both sequential and parallel composition and offers a powerful and extensible specification and proof techniques for reasoning for safety, liveness and projected time [22]. Several temporal logic interpreters based upon ITL are available [13].

An interval with fixed length is defined as $X = [sx, ex]$, where $ex \geq sx$. In this definition sx denotes "start of X" and ex denotes "end of X". The formalism considers a single time line. Any two intervals $X = [sx, ex]$ and $Y = [sy, ey]$ on this time line are related by one of seven relationships defined as follows:

Definition 1: *Set of Point-Interval Temporal Logic Relations.*
Let the set of temporal relations be denoted by R_1:

$$R_1 = \{Before, Meets, Overlaps, Starts, During, Finishes, Equals\}$$

Often, the duration of a task is not known in advance and we can only predicate a lower and an upper bound for its duration. Although the ITL model is suitable for describing deterministic temporal relations, it does not allow modeling and analysis of temporal relations between non-deterministic, or random intervals.

We can also express X as:

$$X = [sx, ex], \ (lx \leq ex \leq ux)$$

This paper proposes a new logic, Extended Interval Temporal Logic, EITL, to handle interval temporal relations when the duration of a task is a random variable with known bounds. When the lower and the upper bound of the termination time of task X coincide, EITL reduces to the ITL model.

To simplify the formalism in this paper we use the same notation for an task (activity), for the transition modeling the task in a TPN model, and for the time interval specifying the duration of the task; the context makes it clear where we refer to a task, a transition, or a time interval.

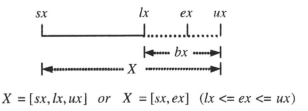

$$X = [sx, lx, ux] \quad or \quad X = [sx, ex] \quad (lx <= ex <= ux)$$

Fig. 4.2 A task X and the extended interval modeling its duration. sx is the starting time, ex is the end time, lx is the lower bound and ux the upper bound on the termination time.

Let X denote a task of indeterminate duration; the starting time of X is denoted as sx, the end is ex; the earliest termination time is denoted lx, the latest termination time of X is ux and the duration from lx to ux is bx (see Figure 4.2). Thus $lx <= ex <= ux$. The length of X is: $|X| = ex - sx$
such that $lx - sx \leq |X| \leq ux - sx$.

Consider two tasks X and Y and assume $sx \leq sy$. Then the relation between the starting time of X and Y could only be one of several temporal relations, see Figure 4.3:

1. X Co-starts Y $(sx = sy)$;
2. X Start-before Y $(sx < sy < lx)$;
3. X Lower-bound-meets Y $(lx = sy)$;
4. X Lower-bound-before Y $(lx < sy < ux)$;
5. X Upper-bound-meets Y $(ux = sy)$;
6. X Upper-bound-before Y $(ux < sy)$.

Definition 2: *Set of Start-point Relations between Two Extended Intervals*

$$R_2 = \{Co - start, Start - before, Lower - bound - meets,$$

57

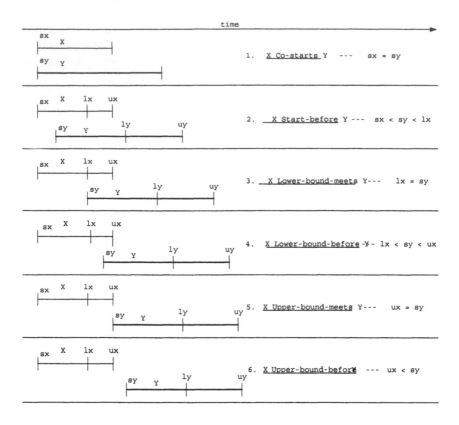

Fig. 4.3 The temporal relationships between two tasks X and Y.

$Lower-bound-before, Upper-bound-meets, Upper-bound-before\}$

Definition 3: *Extended Interval Temporal Logic $R_2 \times R_1$. The temporal relation between two tasks X_i and X_j is a Boolean function:*

$$R_2 \times R_1 \longrightarrow \{T, F\}$$

where T and F are the Boolean values, $T(true)$ and $F(false)$. R_2 denotes the relations between the two start-points of the extended intervals, and R_1 denotes the relations between the two bound intervals. The combination of R_1 and R_2 defines all temporal relations of EITL.

We use a TPN model to show the temporal relationship between two or more tasks and implicitly between the corresponding extended intervals. In the following $X = [sx, lx, ux]$, $Y = [sy, ly, uy]$, $Z = [sz, lz, uz]$ represent three tasks modeled as extended intervals.

Consider the case when the tasks are independent of each other. Then the transitions modeling two tasks, X and Y fire independently, the TPN model is relatively simple, and we can classify the models according to the relation between the starting times of X and Y.

Figure 4.4 illustrates the case when the two tasks have the same starting time.

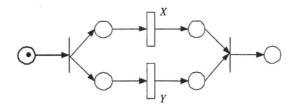

Fig. 4.4 The extended intervals and the TPN model for the X Co-starts Y temporal relationship between tasks X and Y.

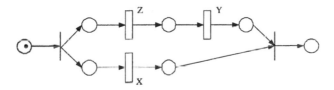

Fig. 4.5 The TPN model for several temporal relationships between tasks X and Y.

Figure 4.5 illustrates several cases involving three tasks, X, Y, and Z such that $sx = sz$ and $sy > uz$. The constraints on Z lead to the following temporal relationships:

- Z Starts-before Y: Z satisfies the condition: $ly > uz$.

- X Low-bound-meets Y: when Z satisfies the condition: $lx = lz = uz$

- X Low-bound-before Y: when Z satisfies the condition: $lx \leq lz < uz < ux$

Figure 4.6 illustrates two cases involving three tasks, X, Y, and Z such that $ux = sz$ and $uz = sy$. The constraints on Z lead to the following temporal relationships:

- X Upper-bound-meets Y. This temporal relation occurs when Z satisfies the condition: $ux = lz = uz$.

59

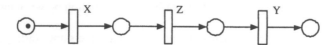

Fig. 4.6 The extended intervals and the TPN model for: X Upper-bound-meets Y and X Upper-bound-before Y temporal relationships between tasks X and Y.

- X Upper-bound-before Y. This temporal relation occurs when Z satisfies the condition: $ux < lz$.

Theorem 1. Given two activities X and Y and a temporal specification from R_2, there exists a TPN representation for the temporal relation between X and Y

Proof: Let the intervals X and Y be $[sx, lx, ux]$ and $[sy, ly, uy]$ respectively, see Figure 4.4. The relation X Low-bound-before Y means that event Y occurs after the lower bound of X, but before the upper bound of X, or $lx < sy < ux$. We want to prove that the TPN representation of the temporal relation in Fig. 4.4(d) with the bound restriction of Z ($lx \leq lz < uz < ux$) satisfies: $lx < sy < ux$.

In Figure 4.5, the transitions X and Z are enabled after the immediate transition fires, thus $sz = sx$. According to the bound restriction of Z, $lx \leq lz < uz < ux$, the end-point of event Z is between the lower and upper bounds of X, i.e., $lx < ez < ux$. Task Y is enabled immediately after Z ends, i.e., $ez = sy$, so Y occurs after the lower bound and before the upper bound of X; thus the relation $lx < sy < ux$ holds. Similarly, we can easily prove that the models in Figure 4.6 represent the corresponding extended interval temporal specifications.

Theorem 1 can be extended for:

- multiple tasks that can be started independently and

- n (n is a positive integer) known statements, each expressing the temporal relationship between two tasks and their extended intervals.

Let us assume that the statements are correlated; this means that we can select k out of the n statements ($1 \leq k \leq n - 1$), such that there exists at least one task which appears in the k statements at least once and also appears on the left hand side of the remaining $n - k$ statements at least once. Then the temporal relations among these tasks can be represented by a connected TPN model.

Theorem 2. Given multiple independent tasks and n correlated temporal statements about them, there exists a connected TPN model representing the relationship among these tasks.

Proof: Select any one of the n statements, statement s. According to Theorem 1 we can build the TPN model for this statement. According to the definition of correlation, we can identify a task X on the left hand side of one of the $n - 1$ statements, statement p which also appears in statement s.

According to the rules, there is a connected TPN model representing statements s and p. Then we attempt to find another statement q from the remaining $n - 2$

60

statements, which must have the same interval with one of the intervals in the first two statements selected above. We combine the TPN model of statement q with the former TPN model and get a new connected TPN model. After $n - 1$ such combinations we get a connected TPN model which represent all n statements.

Now assume that we have a known temporal relationship between two extended intervals X and Z and a new temporal relation between two extended intervals Z and Y is discovered. Then the inference engine based upon the temporal model discussed in this paper can construct an analytical representation of the temporal relation between X and Y with the help of the known relations among X, Y and Z.

The construction of the analytical representation of an unknown relation between two intervals, X and Y, from the known temporal relation among system's other intervals can be accomplished using the following algorithm:

Step 1: Construct the composite TPN model of EITL representation according to the known temporal relations;

Step 2: Simplify the TPN model by deleting the redundant subnets according to the computation of S-invariants;

Step 3: Calculate the relation between X and Y according to the axioms of algebraic inequalities;

Step 4: Convert the temporal relation of points into EITL relations.

In general, an inference engine based on the TPN model has to search all possible transitions related to X and Y to determine the right relation between X and Y. Commonly, a system's TPN description contains redundant information for a specific temporal inference, so the TPN model can be firstly simplified and refined before reasoning.

We now present rules for combining several TPN models of temporal statements regarding the activities of a workflow. The rules involved in this reduction process are:

Rule 1 - Delete immediate transitions.

Rule 2 - Combine the same transitions and places.

Rule 3 - Add new immediate transitions or use already existent transitions to synchronize the known start-points of intervals.

Rule 4 - Add immediate transition to synchronization of end point of activities.

Example 1: Consider the following temporal logic statements related to activities X, Y, and W:

1) X Low-bound-meets Y,

2) Y Low-bound-before W.

The TPN representation of the two statements is given in Figure 4.7(a). Since time transitions X and Z_1 can fire at the same time, and no new synchronized firing interval is added, T_1 will not be deleted according to the Rule 3. T_3 can be deleted because transitions Y and Z_2 can fire at the same time and Z_1 is a time transition, according to Rule 1. Z_1 alone can synchronizes Y and Z_2. Finally, delete T_2, T_4, and add T_5 to realize the synchronization of end points as shown in Figure 4.7(b).

Example 2: Let a workflow containing activities X, Y, W and P be described by the following statements:

(1) X start-before Y, bx overlaps by

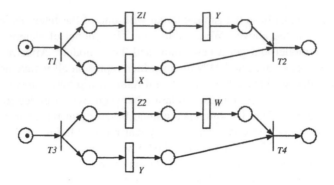

Fig. 4.7 Example 1. (a) The TPN model. (b) The composite TPN model.

(2) Y low-bound-before W, *by* meets bw
(3) Y start-before P, *by* during bp
The task of the inference engine is to infer the temporal relationship between intervals X and P.

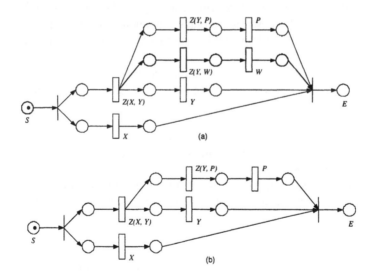

Fig. 4.8 Example 2. (a). The TPN. (b). Simplified model.

Step 1: construct a composite TPN model according to the known temporal relations, see Figure 4.8(a);
Step 2: simplify the TPN model of temporal relations for inference between X and P. By computing the S-invariants of the TPN model, we can get four paths, the time

62

transition s for each path is as follows:

$$\|x_1\| = \{Z(X,Y), Z(Y,P), P\};$$

$$\|x_2\| = \{Z(X,Y), Z(Y,W), W\};$$

$$\|x_3\| = \{Z(X,Y), Y\};$$

$$\|x_4\| = \{X\}.$$

For inference between X and P, it can be seen that the transitions in sets $\|x_1\|$, $\|x_3\|$ and $\|x_4\|$ are closely correlated. So a simplified TPN model can be constructed by the paths associated with $\|x_1\|$, $\|x_3\|$ and $\|x_4\|$, while the path associated with $\|x_2\|$ is redundant and should be deleted. The simplified TPN model of EITL is shown in Figure 4.8(b).

Step 3: Calculate the relation between X and Y according to the axioms of algebraic inequalities. We get the result as below:

$$sy < sp < lp < ly < uy < up \qquad (1)$$

$$sx < sy < lx < ly < ux < uy \qquad (2)$$

combine (2) with (1) delete sy, ly and uy, we get:

$$sx < sp < ux; sx < lp < ux; up > ux. \qquad (3)$$

Step 4: Convert the temporal relation of points into the EITL relations.

As $sx < sp < ux$, the possible start-point relations between X and P are as follows:

X Start-before P $(sx < sp < lx)$
X Low-bound-meets P $(sp = lx)$
X Low-bound-before P $(lx < sp < ux)$

When X Start-before P $(sx < sp < lx)$, combined with $sx < lp < ux < up$, the possible bound interval relations between X and P are below:

bx During bp $(lp < lx < ux < up)$
bx Starts bp $(lp = lx < ux < up)$
bx Overlaps bp $(lx < lp < ux < up)$

When X Low-bound-meets P $(sp = lx)$, combined with $sx < lp < ux < up$, the possible bound interval relations between X and P are below:

bx Starts bp $(lp = lx < ux < up)$
bx Overlaps bp $(lx < lp < ux < up)$

When X Low-bound-before P $(lx < sp < ux)$, combined with $sx < lp < ux < up$, the possible bound interval relations between X and P are below:

bx Overlaps bp $(lx < lp < ux < up)$.

63

4.4 LINEAR REASONING ALGORITHMS FOR BASIC WORKFLOW PATTERNS

A process requires the activation of multiple tasks. We recognize two types of *activity tasks*, those whose activation has side effects and *routing tasks* that determine the flow of control among tasks. Throughout the paper we call the activity task simply tasks and denote them by $X, Y, Z, \ldots$; we use the actual name of a routing task e,g., *XOR split/join*.

The term *workflow pattern* refers to the temporal relationship amongst the tasks of a process. A workflow description languages and the mechanisms to control the enactment of a case must have provisions to support these temporal relationships.

A collection of workflow patterns is available at [4]; the authors of classify these patterns in several categories: basic, advanced branching and synchronization, structural, state-based, cancellation, and patterns involving multiple instances.

We now overview the workflow patterns from Figure 4.9:

(a) The *sequence* pattern occurs when several tasks have to be scheduled one after the completion of the other. In our example task Y can only be started after X has completed its execution and, in turn, Y has to finish before task Z can be activated.

(b) The *AND split* pattern requires several tasks to be executed concurrently. Both tasks Y and Z are activated when task X terminates. In case of an *explicit AND split* the activity graph has a routing node and all activities connected to the routing node are activated as soon as the flow of control reaches the routing node. In the case of an *implicit AND split* activities are connected directly and conditions can be associated with branches linking a task with the next ones. Only when the conditions associated with a branch are true, the tasks are activated.

(c) The *synchronization* pattern require several concurrent activities to terminate before an task can start; in our example, task Z can only start after both tasks X and Y terminate.

(d) The *XOR split* requires a decision; after the completion of task X, either Y, or Z can be activated.

(e) The *XOR join* occurs when several alternatives are merged into one, in our example Z is enabled when either X or Y terminate.

(f) The *OR split* pattern is a construct to choose multiple alternatives out of a set. In our example after completion of X, one could activate either Y, or Z, or both.

(g) The *multiple merge* construct allows multiple activation of a task and does not require synchronization after the execution of concurrent tasks. Once X terminate, tasks Y and Z execute concurrently. When the first of them, say Y, terminates then task W is activated; then when Z terminates, W is activated again.

(h) The *discriminator* pattern waits for a number of incoming branches to complete before activating the subsequent task; then it waits for the remaining branches to finish without taking any action until all of them have terminated. Then it resets itself.

(i) The *N out of M join* construct provides a barrier synchronization. Assuming that $M > N$ tasks run concurrently, N of them have to reach the barrier before the next task is enabled. In our example any two out of the three tasks, X, Y, and Z have to finish before V is enabled.

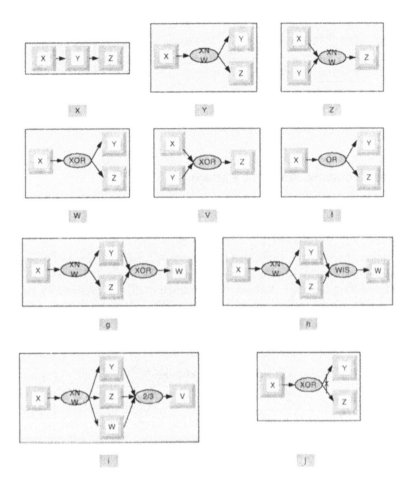

Fig. 4.9 Basic workflow patterns. (a) Sequence; (b) AND-split; (c) Synchronization; (d) XOR-split; (e) XOR-merge; (f) OR-split; (g) Multiple Merge; (h) Discriminator; (i) N out of M join; (j) Deferred Choice.

(j) The *deferred choice* pattern is similar with the *XOR split* but this time the choice is not made explicitly and the run-time environment decides what branch to take.

In the following we analyze several patterns and propose a set of linear reasoning algorithms to reduce each pattern to a canonical form. The original model has a single starting point A and a single end point C. Point A and C are connected by a subnet. After reduction the subnet is substituted by an equivalent transition T, see Figure 4.10.

Fig. 4.10 Reduction to a canonical form requires the replacement of a subnet with an equivalent transition.

4.4.1 The Sequential Workflow Pattern

The following algorithm expresses the process of reduction in case of a sequential workflow pattern:

$Algorithm\ 1\ (T_1, T_2) \xrightarrow{1} T[t_l, t_u] = T[t_{1l} + t_{2l}, t_{1u} + t_{2u}]$

Justification: Figure 4.11 shows a TPN model of a sequential workflow pattern consisting of two tasks X to be executed first and Y to be executed second. Now the subnet modeling the two tasks consists of transitions $T_1[t_{1l}, t_{1u}]$ and $T_2[t_{2l}, t_{2u}]$. The equivalent transition of the canonical form is $T[t_l, t_u]$, with t_l and t_u unknown. Our goal is to relate t_l and t_u with t_{1l}, t_{1u}, t_{2l}, and t_{2u} .

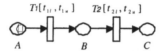

Fig. 4.11 The TPN model of a sequential workflow pattern. The two tasks X and Y are modeled as transitions $T_1[t_{1l}, t_{1u}]$ and $T_2[t_{2l}, t_{2u}]$.

Call s_1, s_2, s are the enabling time and t_1^*, t_2^*, t^* the firing time of T_1, T_2 and T respectively. From the definition of TPN it follows that:

$$t_{1l} \le t_1^* \le t_{1u}; t_{2l} \le t_2^* \le t_{2u}; and\ t_l \le t^* \le t_u$$

From the definition of the Sequential pattern: $s_1 + t_1^* = s_2$;
T is the equivalent transition thus: $s = s_1, t^* = s_2 + t_2^*$;
Form the axioms algebraic inequalities, we get:

$$t_{1l} + t_{2l} \le t^* = t_1^* + t_2^* \le t_{1u} + t_{2u}$$

It follows that:
$t_l = t_{1l} + t_{2l}; and\ t_u = t_{1u} + t_{2u}$.

4.4.2 The AND Split/join Workflow Pattern.

The following algorithm expresses the process of reduction in case of AND split/join patterns:

66

Algorithm 2: $(T_1, T_2) \xrightarrow{2} T[t_l, t_u] = T[max(t_{1l}, t_{2l}), max(t_{1u}, t_{2u})]$

Justification: the TPN model of the *AND split/join* pattern is presented in Figure 4.12. Transitions T_1 and T_2 are executed in parallel; they could fire at the same time, or in any order. This pattern involves two basic flow-control structures, *AND-split* and *AND-join*; the *AND-split* means that T_1 and T_2 can fire simultaneously; the function of the *AND-join* is to synchronize the two sub-flows, and produce a new token in place B after T_1 and T_2 have fired. Both structures are represented by instantaneous transitions or by time transitions.

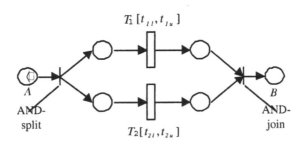

$T_1[t_{1l}, t_{1u}]$

A

AND-split

B

AND-join

$T_2[t_{2l}, t_{2u}]$

Fig. 4.12 The TPN model of a *AND split /join* workflow pattern. The two parallel tasks X and Y are modeled as transitions $T_1[t_{1l}, t_{1u}]$ and $T_2[t_{2l}, t_{2u}]$.

T_1 and T_2 make up a subnet. Call s_1, s_2, s are the enabling time and t_1^*, t_2^*, t^* are the firing time of T_1, T_2 and T respectively. T_1 and T_2 are enabled at the same time, that is $s_1 = s_2 = s_0$.

According to the definition of the TPN: $t_{1l} \leq t_1^* \leq t_{1u}; t_{2l} \leq t_2^* \leq t_{2u};$

That means that the end point of T_1 is between $[s + t_{1l}, s + t_{1u}]$, the end point of T_2 is between $[s + t_{2l}, s + t_{2u}]$, the earliest time both T_1 and T_2 could completed is $max(s + t_{1l}, s + t_{2l})$, and the latest time both T_1 and T_2 could completed is $max(s + t_{1u}, s + t_{2u}$. Thus $max(t_{1l}, t_{2l}) \leq t^* \leq max(t_{1u}, t_{2u})$ and: $t_l = max(t_{1l}, t_{2l}), t_u = max(t_{1u}, t_{2u})$.

4.4.3 The *OR Split/join* Pattern; The Non-deterministic Choice.

The following algorithm expresses the process of reduction in case of an *OR split/join*:

Algorithm 3 $(T_1, T_2) \xrightarrow{3} T[t_l, t_u] = T[min(t_{1l}, t_{2l}), min(t_{1u}, t_{2u})]$

Justification: the TPN model of the *OR split/join* pattern is presented in Figure 4.13. The transitions T_1 and T_2 are enabled at the same time, but they are in conflict: if T_1 fires then T_2 will not fire, and vice versa. The system chooses to fire one of several enabled transitions.

Two basic flow-control structures are used in the model to choose one from two or more transitions. OR-split and OR-join. An OR-split structure is represented by a position, which have several output arcs. An OR-join structure is represented by a position, which have several input arcs. A token set out from position A, after an

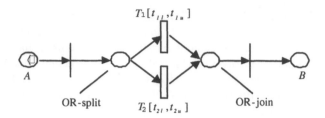

Fig. 4.13 The TPN model of a *OR split /join* workflow pattern. Only one of the two parallel tasks X and Y, modeled as transitions $T_1[t_{1l}, t_{1u}]$ and $T_2[t_{2l}, t_{2u}]$, will ever be executed. The choice is non-deterministic.

instantaneous transition, T_1 or T_2 fires, and then after another instantaneous transition, the token reaches position B.

The transition T fires either according to T_1 or according to T_2. Follow the definition of free selection, the earliest firing time of T is $min(t_{1l}, t_{2l})$ and the latest firing time of T is $min(t_{1u}, t_{2u})$. Thus: $t_l = min(t_{1l}, t_{2l})$, $t_u = min(t_{1u}, t_{2u})$.

If $t_{1u} < t_{2l}$, T_1 will fire inevitably, and T_2 will never fire.

4.4.4 The *XOR Split/join* Pattern; Deterministic Choice.

The following algorithm expresses the process of reduction in case of a deterministic choice:

$Algorithm\ 4$: $(T_1, T_2) \xrightarrow{4} T[t_l, t_u] = T[min(t_{1l}, t_{2l}), max(t_{1u}, t_{2u})]$ $(T_1 \cap T_2 \neq \emptyset)$

$(T_1, T_2) \xrightarrow{4} T[t_l, t_u] = [t_{1l}, t_{1u}] \cup [t_{2l}, t_{2u}]$ $(T_1 \cap T_2 = \emptyset)$

The *XOR split* routing task has a condition c associated with it to model a deterministic choice. One of the two parallel tasks X or Y task is activated, depending upon c, see Figure 4.14(a); if $c > 0$, T_1 fires; if $c \leq 0$, T_2 fires.

The equivalent transition T fires either according to T_1 or according to T_2. The firing interval of T is a combination of $[t_{1l}, t_{1u}]$ and $[t_{2l}, t_{2u}]$. Consider two cases:

(1) when $[t_{1l}, t_{1u}] \cap [t_{2l}, t_{2u}] \neq \emptyset$, the combination of T_1 and T_2 builds up one new continuous interval, with the earliest firing time $min[t_{1l}, t_{2l}]$ and the latest firing time $max[t_{1u}, t_{2u}]$. Thus: $t_l = min(t_{1l}, t_{2l})$, $t_u = max(t_{1u}, t_{2u})$.

(2) when $[t_{1l}, t_{1u}] \cap [t_{2l}, t_{2u}] = \emptyset$, the firing interval of the equivalent transition will be two continuous intervals: $[t_{1l}, t_{1u}] \cup [t_{2l}, t_{2u}]$. Again: $t_l = min(t_{1l}, t_{2l})$, $t_u = max(t_{1u}, t_{2u})$.

4.4.5 The Iteration Pattern.

The following algorithm expresses the process of reduction in case of an iterative pattern, see Figure 4.15 (a).

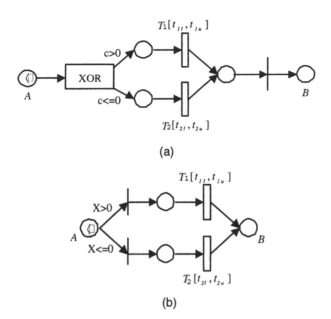

(a)

(b)

Fig. 4.14 The TPN model of a *XOR split /join* workflow pattern. (a) Only one of two parallel tasks X and Y, modeled as transitions $T_1[t_{1l}, t_{1u}]$ and $T_2[t_{2l}, t_{2u}]$, will ever be executed. The choice is deterministic: if $c > 0$, T_1 fires; if $c \leq 0$, T_2 fires. (b) Two conditions are associated with arcs from A.

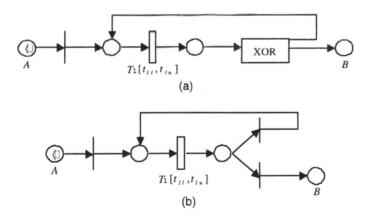

(a)

(b)

Fig. 4.15 The TPN model of an iteration workflow pattern. (a) The original TPN . (b) The transformed TPN model.

Algorithm 5: $(T_1, k) \overset{5}{\longrightarrow} T[t_l, t_u] = T[kt_{1l}, kt_{1u}]$ In Figure 4.15 (a) we see that this the flow-control structure has an XOR routing that tests the results of the task

69

modeled by transition T_1, and based upon this test, T_1 may be executed k times. Figure 4.15 (b) shows the equivalent model without an XOR routing task.

4.5 LINEAR REASONING ON WORKFLOW MODELS.

We now provide a set of minimal requirements for the TPN model to ensure the correctness of the workflow:

1) A TPN model has a *source place i* corresponding to start condition and a *sink place o* , corresponding to an end condition.

2) Each task/condition is on a path from i to o.

These two conditions are related to the structural properties of the Petri net and can be verified using a static analysis; they are similar to the ones imposed to the WF-nets defined by van de Aalst [2]. For linear reasoning we impose two additional conditions:

3) For any case, the procedure will terminate eventually and the moment the procedure terminates there is a token in place o and all the other places are empty.

4) There should be no dead tasks, i.e., it should be possible to execute an arbitrary task by following the appropriate route though the workflow model.

Definition 4 *(Soundness) A procedure modeled by a WF-net* $PN = (P, T, F)$ *is sound iff:*

(1) For every state M reachable from state I, there exists a firing sequence leading from state M to state O. Formally:

$$\forall_M (I \xrightarrow{*} M) \implies (M \xrightarrow{*} O)$$

(2) The state O is the only state reachable from state I with at least one token in place o. Formally:

$$\forall_M (I \xrightarrow{*} M \land M \geq 0) \implies (M = O)$$

(3) There are no dead transitions in the WF-net. Formally:

$$\forall_{t \in T} \exists_{M,M'} I \xrightarrow{*} M \xrightarrow{t} M'$$

Next we use an example from [5] to illustrate the method for establishing temporal relations in a WF-net.

Example 3: We model the activity of a traveling agency and follow a client interested in booking a trip,see Figure 4.16. First, the client waits until a travel agent becomes available and then provides the travel information: the destination, the date, the desired departure time, the type of lodging, rental car information, and so on; then it pays a service fee. This requires 20-30 minutes all together.

Then the travel agent searches for availabilities and after some 60 minutes has a full itinerary and discusses it with the client. There are three possible outcomes of this discussion; the client:

- accepts the itinerary; then the agent contacts the hotel and the airline (10-30 minutes) and confirms the bookings, this takes 20-40 minutes. If the customer

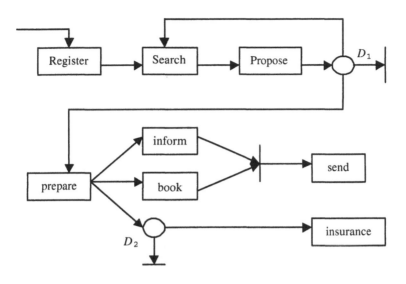

Fig. 4.16 The task structure corresponding for booking a trip at a travel agency.

requires insurance an additional time is requires, 15-30 minutes. Some of these tasks can be executed in parallel, see Figure 4.16.

- asks for changes; the agent spends another 60 minutes changing the reservations.

- leaves without booking; he spends an additional 5-10 minutes to have his service fee refunded.

We want to determine:

(1) The time required when one change is requested but cannot be accommodated and the client leaves without booking the trip.

(2) The shortest possible time a client needs for booking a trip.

We use an equivalent TPN model to represent the process of booking a trip; in Figure 4.17, P_1 is a source place, and P_10 is a sink place.

From P_1 to P_3, the structure is s combination of the sequential and the iteration patterns. Using algorithms 1 and 5, the equivalent transition from P_1 to P_2 called T_1' is:

$$(T_2, k) \xrightarrow{5} T[kt_{1l}, kt_{1u}] = T[60k, 60k]$$
$$(T_1, T[60k, 60k]) \xrightarrow{1} T[20 + 60k, 30 + 60k] = T_1'$$
$$(T_1, T_2) \xrightarrow{1} T[t_l, t_u] = T[t_{1l} + t_{2l}, t_{1u} + t_{2u}]$$

The tasks inform and book can be executed in parallel; using algorithm 2, we obtain the equivalent transition T_2' as follows:

$$(T_5, T_6) \xrightarrow{2} T[max(10, 20), max(30, 40)] = T[20, 40] = T_2'$$

71

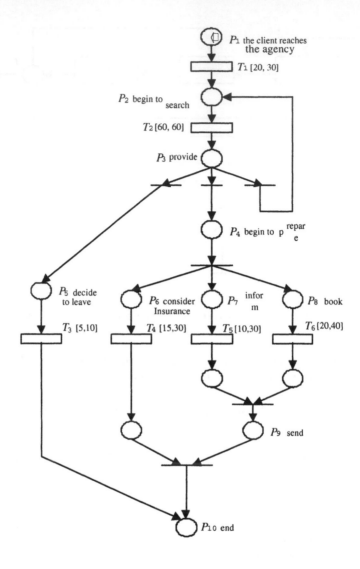

Fig. 4.17 The TPN model for the process in Figure 16, describing the booking of a trip.

The transitions T_2' and T_4 can be executed in parallel; using algorithm 2 get obtain the equivalent transition corresponding to the preparation process T_3':

$$(T_4, T_2') \xrightarrow{2} T[max(15, 20), max(30, 40)] = T[20, 40] = T_3'$$

The transitions T_3 corresponding to "leave without booking" and T_3', to "preparation" are condition selection; using algorithm 4:

$$(T_3, T_3') \xrightarrow{4} T = [5, 10] \cup [20, 40]$$

The equivalent transition corresponding to the case when the client leaves without booking is modeled by transition T_4':

$$(T_1', T[5, 10]) \xrightarrow{1} T[60k + 5, 60k + 10] = T_4'$$

(1) When $k = 2$, the time taken by the equivalent transition T_4' is from 2 hours and 5 minutes, to 2 hours and 10 minutes.

Assume the equivalent transition from the client reaches the agency to he books a trip successfully is transition T_5':

$$(T_1', T[20, 40]) \xrightarrow{1} T[60k + 20, 60k + 40] = T_5'$$

(2) When $k = 1$, the shortest time a client needs for booking the trip is 60+20=80 minutes,.

4.6 CONCLUSIONS

Traditional applications of worflow management to business processes and office automation typically deal with few activities with a relatively long lifetime, hours, days, weeks, or months; the time scale reflects the fact that most of the activities are carried out by humans while the enactment engine is hosted by a computer.

On the other hand, modern applications of workflow management to process coordination on service, data and computational grids deal with a larger number of short-lived activities with a lifespan of microseconds, milliseconds, seconds, or minutes, carried out by computers. While in the first case it is feasible to use ad-hoc methods to study the temporal properties of processes, this option becomes impractical in the second case. Thus need for automated methods to study the temporal behavior of workflow models. Moreover, the second type of applications require very efficient algorithms for workflow enactment and analysis due to the very short nature of activities involved.

In this paper we investigate the temporal properties of workflows modeled by WF-nets. A set of linear reasoning algorithms for commonly used workflow patterns are presented. Using these algorithms one could solve temporal reasoning problems for sound WF-nets in linear time.

The approach discussed in this paper does not take into account the case of random intervals; an extension of the approach will incorporate these features into the methodology. The reasoning algorithms would be extended to answer queries regarding random intervals of an event.

4.7 ACKNOLWEDGMENTS

The work reported in this paper was partially supported by several grants from the National Science Foundation, MCB-9527131, DBI-9986316, and ACI-0296035. The first author acknowledges also the support from CNSF under grant number 60173012, and Chinese Development Plan of the State Key Fundamental Research (973) under grant number G1999032707.

REFERENCES

1. van der Aalst W. M. P. Three Good Reasons for Using a Petri Net based Workflow Management System. Proc. IPIC 96, 179-181, 1996.

2. van der Aalst W. M. P. The Application of Petri Nets to Workflow Management. Journal of Circuits, Systems, and Computers, 8(1): 21-66, 1998.

3. van der Aalst W. M. P. Workflow Verification: Finding Control-Flow Errors Using Petri-Nets-based Techniques. In Business Process Management. W. van der Aalst, J. Desel, and A. Oberweis, Eds. Lecture Notes in Computer Science, Vol. 1806, 161-183, 2000.

4. van der Aalst and A. H. ter Hofstede and B. Kiepuszewski and A.P. Barros. Workflow Patterns. http://www.tm.tue.nl.research/patterns/.Technical Report, Eindhoven University of Technology, 2000.

5. van der Aalst W.M.P. Verification of Workflow Task Structures. Information Systems, 25(1):43-69, 2000.

6. Adam, N.R. V. Atluri, and W. K. Huang. Modeling and Analysis of Workflows Using Petri Nets. Journal of Intelligent Information Systems. 10(2), 1-29, 1998.

7. Allen J. F., Maintaining knowledge about temporal intervals, Communications of ACM, 26(11):832-843, 1983.

8. Allen J.F., Toward a general theory of action and time, Artificial Intelligent, 23(2):123-154, 1984.

9. Bracchi, G. and B. Pernici, The Design Requirements of Office Systems, ACM Trans. Office Automat. Syst., 2(2):151-170, 1984.

10. Chinn, S.J., G. R. Madey. Temporal Representation and Reasoning for Workflows in Engineering Design Change Review. IEEE Trans. On Engineering Management, 47 (4): 485-492, 2000.

11. Ellis, C.A. and G.J. Nutt. Modeling and Enactment of Workflow Systems. In Applications and Theory of Petri Nets, Lecture Notes on Computer Science, Vol. 691, Springer Verlag, 1-16, 1993.

12. Ellis, C.A., K.Keddara, and G. Rozenberg. Dynamic Change within Workflow Systems. In N.Comstock and C.Ellis, Eds., Conf.on Organizational Computing Systems, ACM SIGOIS, ACM, 10-21, Milpitas, CA, 1995.

13. Fujita M., Kono S. and Tanaka H., Tokio: LP Language based on Temporal Logic and its compilation to Prolog, in Proc. Int. Conf. on Logic Programming, London, July 1986.

14. Jensenss, G.K., J. Verelst, B. Weyn. Techniques for Modelling Workflows and their Support of Reuse. In Business Process Management. W. van der Aalst, J.

Desel, and A. Oberweis Eds. Lecture Notes in Computer Science, Vol. 1806, 1-15, 2000.

15. Lin, C. and Chanson S T. Logical inference of clauses based on Petri net models. International Journal of Intelligent Systems? John Wiley & Sons, 13: 821-840, 1998.

16. Lin, C., A. Chaudhury, A.B. Whinston, and D. C. Marinescu. Logical inference of Horn clauses in Petri net models. IEEE Trans. on Knowledge and Data Engineering, 5(4): 416-425, 1993.

17. Marinescu D.C. Internet-based Workflow Management; Towards a Semantic Web. 610+XVII, Wiley, New York, Chichester, Weinheim, Brisbaine, Singapore, Toronto, 2002.

18. Mittasch C., T. Weise, and M. Hesselmann. Decentralized Control Structures for Distributed Workflow Applications. Integrated Computer-Aided Engineering, 7(4): 327-34, 2000.

19. Yi Zhou, Murata T., Fuzzy-timing Petri net model for distributed multimedia synchronization, IEEE International Conference on Systems, Man, and Cybernetics, Vol 1, 244 -249,1998.

20. Moszkowski B., Reasoning about Digital Circuits, Technical Reports STAN-CS-83-970, PhD thesis, Department of Computer Science, Stanford University, 1983.

21. Moszkowski B., A temporal logic for multilevel reasoning about hardware, Computer, 18:10-19, 1985.

22. Moszkowski B., Some very compositional temporal properties, In E.-R. Olderog, editor, Programming Concepts, Methods and Calculi, vol. A-56 of IFIP Transactions, IFIP, Elsevier Science B.V. (North-Holland), 307-326, 1994.

23. Puustjarvi J. Workflow Concurrency Control. Computer Journal, 44 (1): 42-53, 2001.

24. Raposo A. B., L.P. Magalhaes, I.L.M. Ricarte. Petri Nets-Based Coordination Mechanisms for Multi-Workflow Environments. Computer Systems Science And Engineering, 15 (5): 315-326, 2000.

25. Schal, T. Workflow Management Systems for Process Organizations. Springer Verlag, 1996.

26. Workflow Management Coalition. http:www.wfmc.com, 1998.

27. Yao, Y. A Petri net Model for Temporal Knowledge Representation and Reasoning. IEEE Trans. Systems, Man, and Cybernetics, 24(9): 1374-1382, 1994.

28. Zaidi A K. On temporal logic programming using Petri nets. IEEE Trans. on Systems, Man, and Cybernetics, 29(3): 245-254, 1999.

Author(s) affiliation:

- **Chuang Lin**

  Department of Computer Science and Technology
  Tsinghua University
  Beijing 100084, China
  Email: chlin@tsinghua.edu.cn

- **Dan C. Marinescu**

  School of Electrical Engineering & Computer Science
  University of Central Florida
  Orlando, FL 32816-2362, USA
  Email: dcm@cs.ucf.edu

5

A Coordination Model for Secure Collaboration

Anand Tripathi
Tanvir Ahmed
Richa Kumar
Shremattie Jaman

Abstract

The focus of this paper is on specifying a coordination model for building secure distributed collaboration and workflow systems from their high level specifications. We identify here unique requirements of security and coordination in dynamic collaboration environments. We present a role-based model for specifying these requirements. This specification model supports hierarchical structuring of a large collaboration environment using nested activities, which can be created dynamically. We briefly describe how a middleware is used to realize and support a collaboration environment from its specifications, enforcing the required policies for coordination and security.

5.1 INTRODUCTION

In computer supported cooperative work (CSCW) systems, multiple users cooperate and collaborate towards some shared objectives and tasks using shared data and communication channels [6]. The fundamental challenges of a CSCW system include sharing data and coordinating activities among the collaborating participants.

Coordination in a CSCW environment can range from tightly coupled real-time interactions, where users are connected to the system simultaneously and interact through sharing and manipulation of graphical and multimedia objects, to loosely coupled interactions, which are largely characterized by workflow environments where the coordination is coarse-grain.

These challenges warrant an increasing awareness of security issues. With collaboration activities possibly spanning different organizations and multiple users concurrently accessing shared objects, there is a need to maintain confidentiality, consistency, and integrity of the shared data. Often, security and coordination are co-dependent in a collaboration environment.

Our goal is to provide a specification language that can express the coordination and security requirements for a range of collaboration systems. From this specification, we can rapidly construct the runtime environment for the system by means of a generic policy-driven middleware [23]. In our model, a policy-driven collaboration system is realized in three steps. Initially, the coordination and security policy for a collaboration is specified based on a schema. From the specification, various policy modules are derived for different kinds of requirements, such as role based security, object level access control, and event notification for coordination. Finally, through these modules, the collaboration environment is realized by a generic middleware.

A policy-driven system is able to specify the dynamic nature of CSCW activities and enforce coordination and security policies at runtime. Existing policy driven CSCW systems, such as COCA [12] and DCWPL [2], are built mainly for shared interactive groupware applications. These systems are for time-space limited activities, and therefore the models for describing the desired systems cannot implement large workflow like CSCW environments where multiple CSCW applications may come into existence within a single collaboration framework.

Our model centers around role-based policies for security and coordination [16, 13]. This model is often used in collaboration systems where participants perform a set of well-defined tasks pertaining to their qualifications and responsibilities in the organization. All privileges are granted to a role rather than a user. A role-based model allows a role to exist independent of its members and provides a means for associating privileges for task execution with the various participants in the system.

A distributed collaboration has a number of coordination and security requirements. These include

1. Hierarchical activity organization
2. Dynamic security model
3. Role management
4. Coordination model

Hierarchical activity organization: A large collaborative environment may sometimes need to be structured hierarchically, consisting of smaller activities nested inside it. For example, a system supporting a team based design project running over several weeks or months may need to have several dynamically created "whiteboard-like" activities of relatively shorter life-spans. Therefore, a facility is needed to hierarchically structure a large CSCW environment into nested sub-activities. Our specification model supports hierarchical and dynamic structuring of a collaboration system using the notion of an *activity*, which defines a protection domain and scoping facility, encapsulating a set of objects, roles, and policies for security and coordination. An activity can have nested sub-activities and multiple instances of the same activity can exist concurrently.

Dynamic security model: The traditional concerns of confidentiality and integrity of shared data are naturally present in collaboration systems. However, access rights, privileges, and ownership of objects may change in collaborative environments as activities progress. Access control may need to be history-based, where the privileges depend on past events. The privileges assigned to a user in a role may change with time due to the actions executed by other participants. Sometimes permissions may change due to the user's own actions. Dynamic access control policies also have to address the issues of role invalidation, role activation, or other conditions that may occur in the system. Privacy can also become an issue when one may need to hide the identity of one participant from another. In such cases, the presence of a participant may only be visible through his/her role or a pseudonym in a role but not by name. The principle of "separation of duties" is an important requirement in a business or office environment [17, 18, 21]. One individual should not be allowed to perform all critical functions in a business transaction. Instead, such responsibilities should be shared among different users to eliminate conflict-of-interest.

Role management: A role represents a set of privileges and can be viewed as a characterization of a protection domain. In role-based systems, static assignment of participants to roles can be error prone and such assignments do not scale well. In a large collaboration environment there is a need for dynamic assignment and delegation of roles. *Role admission constraints* and *activation constraints* address this issue and are described in Section 5.3. Participants may join/leave a role or alternatively be admitted/removed from roles. During the lifetime of a role, a list of the current members of the role needs to be maintained. As participants join or are admitted to a role, various checks need to be performed to conform to the security policies. Similar checks need to be done when any task is performed by a participant in a role.

Coordination model: Coordination among participants in different roles within an activity is based on events generated by the actions of the collaborating participants. For some roles, multiple users may be allowed to simultaneously acquire and activate the role. Coordination among such users can be referred to as *intra-role* coordination. Participation of multiple users in a role can be *independent* or *cooperative*. In independent participation, every member assumes responsibility for certain role-specific tasks, irrespective of the presence of other participants. For example, in a document authoring activity having three members in a *Reviewer* role, each member writes an independent review. When participants coordinate and share the tasks of the role, it is referred to as *cooperative* participation. For example, in a medical system, certain procedures might be required to be performed on a patient by any one nurse. Another form of intra-role coordination requires that all users in the role must participate for a certain operation to be successful. An example would be a group of participants jointly opening a bank vault. Some scenarios may also war-

79

rant less rigid coordination models, e.g. in an unrestricted whiteboard sharing activity.

This paper presents a specification model which addresses the above security and coordination requirements within hierarchically nested collaborative systems and which facilitates the derivation of policy modules by a generic middleware to support the runtime environment.

Based on this model, we have devised an XML schema, which facilitates collaboration specifications. We have developed a middleware that implements the XML based policy specifications to realize a collaboration environment. This approach allows one to easily install and experiment with new policies. It also allows us to perform some static checks for consistency in the specifications.

An overview of our collaboration model is presented in Section 5.2 with an example specification of a collaboration environment. Section 5.3 describes how roles, objects and activities are specified in our model. An overview of the middleware execution model is discussed in Section 5.4. Sections 5.5 and 5.6 discuss the related work and the conclusions.

5.2 OVERVIEW OF THE COLLABORATION MODEL

In our model, an *activity template* specifies a generic pattern for collaboration among a set of roles using some shared objects. Any number of instances of it can be dynamically and concurrently instantiated. There are three major elements in specifying a collaboration activity: shared objects, roles, and operations. These entities are specified and named in nested scopes of activities as shown in Figure 5.1. Every activity instance has two meta roles: *Creator* and *Owner*. The user instantiating the activity becomes a member of the *Creator* role. However the owner of the activity may be different from its creator and can be specified in the activity's definition. If not specified, the owner of its parent activity becomes the default owner. The owner of an activity also owns the encapsulated roles, and is trusted for the management of that activity and its nested entities.

5.2.1 Shared Objects

Shared objects are managed by a set of trusted object servers, which are responsible for protecting the objects according to security policies. Each activity can designate an object server to be used for storing all the objects created within the scope of the activity. Based on the security and coordination requirements specified in a collaboration, our system derives appropriate policy modules, similar to the adapter approach [22], which are used by the servers to control access to their objects. The server is also responsible for managing recovery and persistent storage of its objects. Caching and replication however are implementation issues and are transparent at the specification level.

The object servers in our system support both role based access control as well as traditional discretionary access control (DAC). Access control lists are maintained based on user-ids as well as role names. Access control policies can be specified for each method of an object or the object itself. Negative access rights (i.e., constraints) enable easier coarse-grain access control specification.

5.2.2 Role Management

In defining a role, we consider the constraints addressed in Section 5.1. These constraints are of three types: the *role admission condition* puts constraints on admitting a participant to a role, the *role activation condition* needs to be satisfied when a role is activated, and the *precondition* of an operation is required to be satisfied for a member in the role to execute it. The owner of a role can admit users to it, subject to the admission constraints; it can also remove an existing participant. The constraint specifications can include event counters. We present details of these constraint specifications in the next section and show how the security requirements discussed in the previous section can be realized.

The specification for a role in an activity can refer to the objects and other roles in that activity. A nested activity may need to have access to the objects in the scope of its parent activity, or a role in the parent activity may need to be assigned to a role in a nested activity. For this purpose, an activity definition needs mechanisms for role assignment and passing of object references as parameters to an activity instance. A role in the parent activity can be assigned to a role in the child activity in the activity's definition. We refer to it as *role reflection*, which means that all the members of the parent role implicitly become members of the role in the child activity. Removal of a participant from the reflected role (i.e. a role in the parent activity), also implies removal from the role in the child activity. A role reflection can also be specified at the time of activity instantiation. A participant in the reflected roles gains expanded privileges comprising of the operations of the child activity role. Moreover, the child activity role can have operations that access the objects in the scope of the reflected role. However, any participant of the reflected role has to comply with the child role's admission constraints.

An operation in our model represents a task of a role. An operation may consist of the execution of a method of a shared object, a synchronization action, or a role request. A role request is an operation either to *leave* the current role or to *join* another role. Coordination policies are specified as preconditions for operations using an event based model [15].

In the specification model several functions are defined for a role. A boolean function *member(role, user)* checks if a participant is present in a role. The function *members(role)* gives the list of participants in a role. We provide a count operator # on lists. Hence, a count of the participants admitted in a role is given by *#(members(role))*.

5.2.3 Events

Events signify state changes in a collaboration environment. Events and event counters are used in our model for specifying coordination and dynamic security policies.

Event types are based on conditions related to different kinds of entities such as activities, roles, operations, and objects. For example, instantiation of an activity, execution of a role operation, admission of a user in a given role etc. represent different type of events. An event can be identified using the scope based naming. Each event is associated with some collaboration object or some meta object, like a role or activity instance. All events are derived from the base event class, which contains certain common attributes such as *type, id, creation-time, activity-id* etc.

5.2.4 An Example of a Collaboration Activity

The example collaboration environment considered in this paper supports the activities of a group of people jointly developing a document. This will serve as a detailed example to illustrate how policy modules are derived from the specification. Through this example, we highlight some of the coordination and security requirements that were described in Section 5.1 and provide mechanisms for expressing such requirements through our specification model. Figure 5.1 introduces the overall model of our authoring example specified as a *DocumentAuthoring* activity template. The figure shows an instance *document* of this template. The template encapsulates the central entities of the collaboration: a shared document, roles, and collaboration constraints. Authoring of the document object composed of multiple chapter objects is the objective of this collaboration. Three roles are defined for the authoring activity: *Author*, *ExternalReviewer*, and *Supervisor*. In the instance of this *DocumentAuthoring* activity template illustrated in the example, the *Supervisor* role can be referred to by its fully qualified name *DocumentAuthoring.document.Supervisor*.

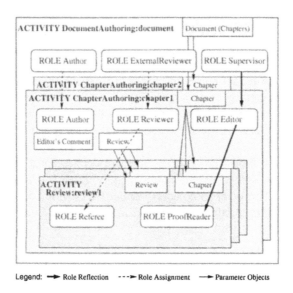

Legend: ➔ Role Reflection ---➤ Role Assignment ➔ Parameter Objects

Fig. 5.1 Hierarchical Structuring of Collaborative Activities

82

In the *DocumentAuthoring* activity, the *ChapterAuthoring* activity is defined as a nested activity with three roles: *Author, Reviewer,* and *Editor.* The document can be composed of any number of chapter objects, and an instance of the chapter object is passed to a *ChapterAuthoring* activity. The example in Figure 5.1 shows two instances of this template: *chapter1* and *chapter2*. A chapter authoring activity manipulates three objects that are shared: the chapter's content, the reviewer's review, and the editor's comments. This *ChapterAuthoring* can be represented as a task-flow diagram, shown in Figure 5.2. Within the *ChapterAuthoring* activity, a participant in the *Author* role writes the contents of the chapter and publishes them. A participant in the *Reviewer* role can then create a *Review* activity in order to prepare an independent review. Two roles, *Referee* and *ProofReader,* are defined for the *Review* activity. When all the reviewers have made their reviews available to the editor, the editor publishes his/her comments, which can be read by the author. Based on the editor's comments, the authors may modify the chapter's content and publish the final version.

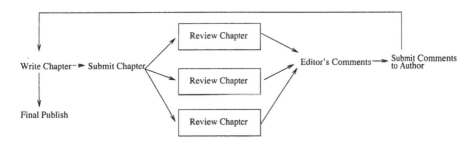

Fig. 5.2 Task flow diagram of ChapterAuthoring Activity

In our example, a participant cannot simultaneously be in the *Author* and the *ExternalReviewer* roles of the *DocumentAuthoring* activity. The *Supervisor* role of the *DocumentAuthoring* activity is *reflected* in the *Editor* role of the *ChapterAuthoring* activity. When an instance of the *ChapterAuthoring* template is created, e.g. *chapter1*, the members of the *Supervisor* role who comply with the *Editor* role's admission constraints are admitted to this role. Therefore, an admitted participant in the *Editor* role can perform operations on the *Document* object in the *DocumentAuthoring* activity as well as on the objects in the *ChapterAuthoring* activity. Similarly, the participants in the *Editor* role are reflected in the *ProofReader* role of the nested *Review* activity. A participant in the *Author* role of the *ChapterAuthoring* activity must be a member of the *Author* role of the parent activity. Similarly, there are certain constraints on the membership of the *Reviewer* role. There may be at most three members in this role, with one member being an author in the parent *DocumentAuthoring* activity but not in the current instance of the *ChapterAuthoring* activity, and the other two being external reviewers. The editor can publish his/her comments only after the chapter's contents have been independently reviewed by three reviewers.

5.3 SPECIFICATION MODEL

5.3.1 Event Specification

Related to each operation are two types of events: *start,* and *finish.* For example, *role-name.op.start* or *role-name.op.finish.* These event types are also defined for each object method. The corresponding event counters can be used for synchronization. For a role, we have event types defined for *join, leave, admit,* and *remove* operations.

Multiple occurrences of a given event type, such as multiple executions of an operation, are represented by a list. The expression *(eventName)* returns the list including all the instances of this type of event. Hence, *#(eventName)* returns the number of times the event has occurred.

In our model, one can also specify a derived event type by filtering an event list based on its attributes. For example, for a role operation execution, we can define a filter based on invoker id, such as opName.start(invoker=John). We can count how many times a user has invoked an operation using #(opName.start(invoker=John)). An example of a role related event is DocumentAuthoring.Author.join(user=Mary).

5.3.2 Object Specification

Shared objects are represented in our specification model only in terms of their types and method signatures, keeping the semantics and implementation details transparent. The method signatures are used in specifying the operations associated with a role. A specification for an object includes a type name to facilitate parameter bindings of operations in roles, a codebase to load the class of the object, and method signatures.

There are three distinct kinds of objects used in our specification: regular objects utilized in a collaborative workspace, roles, and activities. We can specify access control at the granularity of the methods invoked on these objects. User level access control is specified in the scope of a role as part of operation specification. The object level access control specification is derived from the role's privileges on objects. After an object is created the owner can specify additional object level access control. Given below is the specification for the *Chapter* object type in our example. Definition for *ContentType* is not shown.

```
OBJECT_TYPE Chapter (CODEBASE=http://codeserver) {
   METHOD writeContent {PARAM ContentType}
   METHOD readContent {RETURN ContentType} }
```

5.3.3 Role Specification

A role is defined in the scope of an activity and can manipulate the objects in that activity. A role specification includes role name, reflected roles if any, role admission constraints, role activation constraints, and role operations with their preconditions.

The basic terms of a role specification are shown below, where [ ] encloses optional terms.

```
ROLE name ([REFLECT role]) {
    [ADMISSION_CONSTRAINTS condition]
    [ACTIVATION_CONSTRAINTS condition]
    [OPERATION name
        [PRECONDITION condition]
        [ACTION method_invocation]]}
```

5.3.3.1 Role Admission Constraints Role admission constraints are enforced when a participant is assigned or admitted to the role. These constraints enforce several "separation of duty" policies, specifically *static separation of duties, user-user conflict, user-role conflict*, and *history-based* conditions [18, 14].

In our example, in order to ensure that no participant is present in both the *Author* and the *ExternalReviewer* role of the *DocumentAuthoring* activity, the *ExternalReviewer* role has an admission constraint that a participant in this role cannot be a member of the *Author* role within the same activity. This is specified as follows:

```
ROLE ExternalReviewer {
  ADMISSION_CONSTRAINTS
    !member(thisUser, Author)
}
```

Similarly, the *Author* role has an admission constraint that specifies that the participant should not be a member of the *ExternalReviewer* role. These two together form an example of a static "separation of duties".

Role admission constraints can take various forms, for example, a list of valid or invalid users, role operation based conditions, prerequisite roles, or cardinality of the role specifying maximum member count.

5.3.3.2 Role Activation Constraints Role activation constraints are enforced when a user invokes a role operation. In contrast, role admission constraints are checked only when a user is admitted to a role. Dynamic "separation of duty" constraints can be specified as part of role activation constraints. For example, the static "separation of duties" requirement for the *Author* and *ExternalReviewer* roles in the *DocumentAuthoring* activity, which is specified by means of admission constraints in Section 5.3.3.1, can be changed to be enforced dynamically using activation constraints as follows:

```
ROLE ExternalReviewer {
  ADMISSION_CONSTRAINTS
    !member(thisUser, Author)
  ACTIVATION_CONSTRAINTS
    !member(thisUser, Author)
    .....
```

}

Assuming that, the *Author* role does not have any admission constraints, a participant who is not currently a member of the *Author* role can first join the *ExternalReviewer* role, satisfying its admission constraints, and later join the *Author* role. However, the activation constraints of the *ExternalReviewer* role ensure that whenever a member of this role invokes a role operation, the invoker is not a member of the *Author* role. In this way "separation of duties" is enforced dynamically.

Activation constraints may specify a minimum and a maximum number of participants to be present in a role. Additionally, activation constraints may specify some coordination requirements, for example, the activation constraints for the *Reviewer* role in the *ChapterAuthoring* activity, as shown in Figure 5.4, ensure that the reviewer cannot perform any operations unless the chapter has been submitted by the author.

5.3.3.3 *Operation Specification*   An operation is specified by a name, an optional precondition, and an action which invokes predefined methods of some objects in the shared workspace. A keyword *new* is reserved for specifying object and activity creation.

Following is an example of an operation specification for the *Author* role in the *DocumentAuthoring* activity.

```
ROLE Author {
  OPERATION startChapterAuthoring {
    ACTION {chapter=new OBJECT(Chapter)
            act=new ACTIVITY ChapterAuthoring (
            chapter, Author=thisUser) }}
}
```

This specification does not include a precondition, but simply specifies the action that must be taken when the operation is performed. The action includes the creation of a new *ChapterAuthoring* activity, passing the shared *Chapter* object as well as assigning the current user to the *Author* role within the nested activity. In this way nested activities can be dynamically instantiated by performing role operations.

When a precondition is specified for an operation, it must be true when the operation is invoked. This allows us to specify a number of coordination and security constraints like *operational separation of duty* and *object based separation of duty*.

For example, in an office system, a manager may prepare an invoice and approve an invoice, but should not be able to approve his/her own invoice. This specification is shown below:

```
OPERATION ApproveInvoice
  PRECONDITION
    #(PrepareInvoice.finish(invoker=thisUser))=0
```

Preconditions and activation constraints also help in specifying coordination policies among collaborating participants. As described in Section 5.1, coordination between the collaboration participants can be of different types. The following is an example of an inter-role coordination. In the *ChapterAuthoring* activity described in Figure 5.1, the *Editor* role has the following activation constraint.

```
ROLE Editor {
  ACTIVATION_CONSTRAINT
    #(Reviewer.StartReview.finish) = 3
}
```

This condition, along with the precondition for the *StartReview* operation of the *Reviewer* role, ensures that the editor cannot perform any operations until the chapter's contents have been reviewed independently by three reviewers. Similarly, Figure 5.4 also illustrates two forms of intra-role coordination. The precondition for the *Write* operation of the *Author* role specifies that only one author can write the contents of the chapter during one iteration.

```
OPERATION Write
  PRECONDITION
    #(Write.start)-#(SubmitChapter.finish)=0
    & #(SubmitChapter.start)-
    #(Editor.SubmitComment.finish)=0
```

This is a form of *cooperative* participation where the members of the role share the responsibilities of that role. *Independent* participation is exhibited in the *Reviewer* role's *StartReview* operation. The precondition for this operation, #(StartReview.start(invoker=thisUser))=0, states that each reviewer can review the chapter contents only once. This constraint also implies that the reviewers write independent reviews. If the constraint is not based on the invoker's id, then only one of the reviewers can write a review for that chapter.

In the example of the *ChapterAuthoring* activity, the preconditions for the various operations of the *Author* role serve to provide a strict order for the sequence of operations in the activity. The preconditions and activation constraints enforce the desired task-flow of the *ChapterAuthoring* activity as described in Section 5.2.

5.3.4 Activity Specification

An activity definition contains the specification of shared objects, roles, ownership, static assignment of users to these roles, references to objects of other activities, and conditions which will terminate the activity. In Figure 5.3, we describe a specification of the *DocumentAuthoring* activity as shown in Figure 5.1.

The nested *ChapterAuthoring* and *Review* activities are shown separately in Figures 5.4 and 5.5, respectively. In Figure 5.3, participants are assigned to the *Author*, *ExternalReviewer*, and *Supervisor* roles when an activity of this template is instantiated, as specified by the ASSIGNED_ROLES tag in the activity declaration. Also, a *Document* object with multiple *Chapter* objects is passed to this activity.

87

```
ACTIVITY DocumentAuthoring (OBJECTS (Document doc),
                           ASSIGNED_ROLES Author, ExternalReviewer, Supervisor) {
   OBJECT_TYPE Chapter (CODEBASE=http://codeserver) {
      METHOD writeContent {PARAM Content}
      METHOD readContent {RETURN Content}
   }
   ROLE Author {
      ADMISSION_CONSTRAINTS
         !member(thisUser, ExternalReviewer)
      OPERATION startChapterAuthoring {
         ACTION {chapter=doc.getChapter()
                 new ACTIVITY ChapterAuthoring(chapter, Author=thisUser)}}
   }
   ROLE ExternalReviewer {
      ADMISSION_CONSTRAINTS
         !member(thisUser, Author)
   }
   ROLE Supervisor {
      ADMISSION_CONSTRAINTS
         #members(thisRole)<=2
   }
   ACTIVITY ChapterAuthoring {
      ......
      ACTIVITY Review { ... }
   }
}
```

Fig. 5.3 Specification of Collaborative Document Authoring Activity

Participants in the *Author* role can instantiate any number of *ChapterAuthoring* activities with *Chapter* objects. The participant, who instantiates the nested activity, is assigned to the *ChapterAuthoring* activity's *Author* role as specified. The *Supervisor* role specifies a cardinality constraint for role admission.

As shown in Figure 5.4, the owner of an instance of the *ChapterAuthoring* activity is the *Supervisor* role of the parent *DocumentAuthoring* activity. During instantiation, an object of type *Chapter* has to be passed to this activity, and members of the *Author* role of this activity have to be assigned. The definition for this activity includes the specification of two object types: *Review* and *Comment*.

The *Reviewer* role of the *ChapterAuthoring* activity has admission constraints composed of multiple conditions. The first three ensure that only one member of the parent *DocumentAuthoring* activity's *Author* role, who is not a member of the current activity's *Author* role can be in this role. The remaining admission constraints guarantee that, only two external reviewers can join this role.

When the author of a chapter completes a draft of the chapter's content, it is published to the chapter's reviewer and editor by invoking the *SubmitChapter* operation. This results in the coordination actions of making the chapter's content available to the participants in the *Reviewer* and the *Editor* roles. This operation also enables a reviewer to write a review of the chapter. The reviewer of the chapter cannot compose the review until the chapter content has been written and submitted. Similarly, the editor cannot write the comments until all reviewers have reviewed the contents of the chapter independently. After the editor publishes the comments, the author may modify the chapter contents and either publish the final version or simply submit the

```
ACTIVITY ChapterAuthoring (OWNER parentActivity.Supervisor,
                          OBJECTS (DocumentAuthoring.Chapter chapter),
                          ASSIGNED_ROLES Author) {
    TERMINATION_CONDITION #(Author.FinalPublish.finish) > 0
    OBJECT_TYPE Review (CODEBASE=http://codeserver) {
        METHOD writeReview {PARAM ReviewType}
        METHOD readReview {RETURN ReviewType}
    }
    OBJECT_TYPE Comment (CODEBASE=http://codeserver) {
        METHOD writeComment {PARAM CommentType}
        METHOD readComment {RETURN CommentType}
    }
    ROLE Author {
        ADMISSION_CONSTRAINTS
            member(thisUser, parentActivity.Author)
        OPERATION WriteChapter{
            PRECONDITION #(WriteChapter.start) − #(SubmitChapter.finish)=0
                         & #(SubmitChapter.start) − #(Editor.SubmitComment.finish)=0}
            ACTION chapter.writeContent(content)}
        OPERATION SubmitChapter {
            PRECONDITION #(Write.finish) − #(SubmitChapter.start)>0}
        OPERATION ReadComment {
            PRECONDITION #(Editor.SubmitComment.finish)>0
            ACTION comment.readComment()}
        OPERATION FinalPublish {
            PRECONDITION #(ReadComment.start)>0}
    }
    ROLE Reviewer () {
        ADMISSION_CONSTRAINTS
            (member(thisUser, parentActivity.Author)
            & !member(thisUser, Author)
            & #(members(thisRole) ∩ members(parentActivity.Author))<1)
            ||((member(thisUser, parentActivity.ExternalReviewer)
            & #(members(thisRole) ∩ members(parentActivity.ExternalReviewer))<2))
        ACTIVATION_CONSTRAINTS
            #(Author.SubmitChapter.finish) >0
        OPERATION StartReview {
            PRECONDITION #(StartReview.start(Invoker=thisUser))=0
            ACTION { review=new OBJECT(Review)
                     new ACTIVITY Review(chapter, review) }}
    }
    ROLE Editor (REFLECT parentActivity.Editor) {
        ADMISSION_CONSTRAINTS
            member(thisUser, parentActivity.Editor)
        ACTIVATION_CONSTRAINTS
            #(Reviewer.StartReview.finish)=3
        OPERATION CreateComment {
            PRECONDITION #(CreateComment)=0
            ACTION { comment=new OBJECT(Comment) }}
        OPERATION WriteComment {
            PRECONDITION #(CreateComment) > 0
            ACTION { comment.writeComment(data) }}
        OPERATION SubmitComment {
            PRECONDITION #(Author.SubmitChapter.finish) − #(SubmitComment.start) > 0
                         & #(WriteComment.finish) > 0 }
    }
    ACTIVITY Review { ... }
}
```

Fig. 5.4   Specification of Chapter Authoring Activity

chapter again. In this case, no new review activities will take place, only the editor
will be able to publish new comments. This cycle goes on until the author invokes

89

```
ACTIVITY Review(OWNER parentActivity.Editor,
                OBJECTS (DocumentAuthoring.Chapter chapter,
                         DocumentAuthoring.Review review)) {
  ROLE Referee {
    ADMISSION_CONSTRAINTS
      member(thisUser, parentActivity.Reviewer)
      & thisActivity.creator=thisUser
      & #members(thisRole)<1
    OPERATION Read {
      ACTION chapter.readContent()}
    OPERATION Write {
      ACTION review.writeReview(data)}
    OPERATION Submit {
      PRECONDITION #(Write.finish)>0}
  }
  ROLE ProofReader (REFLECT parentActivity.Editor) {
    ACTIVATION_CONSTRAINTS
      #(Referee.Submit.finish)>0
    OPERATION ReadReview {
      ACTION review.readReview()}
  }
}
```

Fig. 5.5 Specification of Review Activity

the *FinalPublish* operation, after which the contents of the chapter are not editable and the *ChapterAuthoring* activity terminates.

This example shows how coordination can be specified in nested activities and how shared objects move among these activities.

5.4 MIDDLEWARE EXECUTION MODEL

Here, we briefly present the middleware execution model to discuss how policy modules are derived and enforced. The middleware provides services for role management, object management, naming, and secure communication of events. A detail discussion of the execution model is presented in [23].

Initially, policy for a collaboration environment is specified using an activity template by the *Convener*. From the activity template, a tree structure is created that represents the various entities in the activity. This structure is built from the XML specification of the collaborative activity and contains only the static definition of the activity. This tree structure is essentially an XML Document Object Model (DOM tree). Figure 5.6 shows the typical structure for realizing a distributed collaboration using a generic coordination facility. As activities are created at runtime, an instance tree is built on the lines of the earlier DOM structure, to represent the distributed runtime environment. Each node in the tree represents an activity, role or an object. Every participant in the system obtains a subset of this tree based on the access control policies. This subset is referred to as the *role specific view* and is accessible to each participant through his/her *User Coordination Interface* (or UCI).

Access control modules are derived from the activity template which specifies the privileges that various roles have on shared objects. In other words, for each object

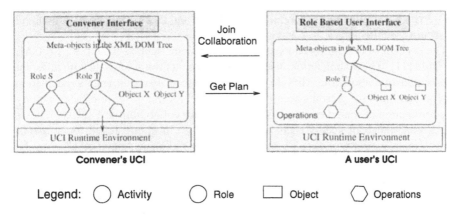

Fig. 5.6 System level view of a distributed collaboration environment

type within the activity, there is a template that contains a skeleton of its access control policies. When these objects get instantiated, the access control entry for the newly created objects get filled in with object instance ids and role ids that are currently active in the activity and are relevant to the objects.

With our specification model, coordination between the participants in a collaboration system is specified in terms of events and event counts. A secure distributed event service, which is a part of the middleware services, derives the subscription-notification relationships between the participants and their operations, from the specification of the activity. Before any activity is created, this subscription-notification model is based on the templates of the various activities, roles and objects that may be created at runtime. As these entities are instantiated at execution time, the parameterized values in templates get bound and the policy modules are accordingly updated.

5.5 RELATED WORK

Our work is similar to COCA [12] and DCWPL [2] in their approach of constructing a distributed collaboration environment from a high level specification. COCA [12] is a logic-based coordination policy specification language for interactive CSCW applications which views security policy as an integral part of coordination policy. At the implementation level, COCA is strongly tied with IP multicast models and Prolog which may not scale for a large collaboration. DCWPL [2] addresses user level mechanism to deal with group interaction issues and is limited to its predefined policies and functions. Neither of these approaches support active security policies, like policy for dynamic separation-of-duty constraints. These approaches can only specify a coordination policy based on previously executed operation and cannot independently specify discretionary access control for an object. Theoretical work for specification of confidentiality in CSCW systems using Z notation is examined in [7], which is limited to its theoretical foundation and lacks implementation model for security architecture.

The concept of role has been used in CSCW systems such as XCP [19], MPCAL [8], Quilt [11]. These applications use roles to represent groups of users with different tasks within a collaboration. They do not satisfy the security requirements specified in Section 5.1. Suite [4] presents access-control model for multiuser interfaces, mainly for coordination of shared editing-based synchronous collaboration. For that, it deals with a wide-range of access rights on shared data. In contrast, the focus of our work is mainly on a participant's role-based access to shared data, addressing the needs of data confidentiality, integrity, and dynamic security policy. In Intermezzo [5], which is designed for user-presence awareness environments, roles are dynamic groups of users with whom access control polices are associated. The use of roles in a collaborative software development environment is presented in [3]. Both of these [5, 3] lack specification of tasks associated with user roles and policies for their coordination.

A task-based constraint specification language for workflow management systems is discussed in [1]. There, constraints are specified with a mapping between roles and tasks. However, it is not clear how such high level specifications of collaborative activities are realized in an implementation. In contrast, our work specifies both the security and the coordination policies, with realization of such policies in an implemented system. SecureFlow [10] imposes workflow authorization constraints on tasks using Authorization Template(AT), which is a tuple specifying privileges to be granted to a subject of a given role on a object of a given type during a given time interval. There, the permissions are activated based on tasks. In contrast, an activity in our model has multiple roles and object types, and the interaction among these roles and objects through operations. If one thinks AT as a method specification, an activity is a module specification with multiple methods. An activity specification may contain multiple tasks or operations and is able to capture workflow stages. For ease of policy specification in distributed systems, the domain concept is used in [24]. However, their domain is a grouping of various objects based on physical location, types, responsibility etc. for the convenience of a policy manager. An activity specification defines a protection domain for several types of objects and roles based on the interaction among these objects and roles.

In [20], team based access control (TMAC) is presented as an approach of applying RBAC in workflow like collaborative environments. In conjunction with RBAC, TMAC proposes an active security model similar to trigger oriented active databases. A team has participants from different roles. Our concept of activity, where roles are created or reflected inside an instance of an activity, subsumes the team concept.

A framework for role based access control (RBAC) models with constraints and role hierarchies is presented in [16]. In [13], role based management (RBM) is presented, which incorporates role obligations, i.e, actions which must be performed when certain events occur, and it views role definitions similar to classes, which can be instantiated or reused through inheritance. In our system, roles are defined as instantiable classes in the context of an activity definition. Instances of the same role in different instances of an activity represent different protection domains. Our role admission criteria are conceptually similar to those presented in the context of RDL (Role Definition Language) in [9]. The focus of that work is on distributed and decentralized management of roles in a distributed service model. In our model, we are concerned with an integrated centralized specification of collaboration environments. Our goal is also to derive an implementation from a specification. We introduce here, the concept role reflection which provides similar functionality of role inheritance and reusability as addressed in [13].

5.6 CONCLUSION

We have presented in this paper a coordination model for secure collaboration systems. We use this model in a policy-driven middleware for supporting rapid construction of a collaboration environment from its specifications. The specification model developed in the context of this work supports nested activities and role based coordination and security policies. In our approach, policy modules are derived from the specifications of a collaboration activity. The policy modules are related to the management of activities, roles, and objects in the collaboration. These policy modules are distributed to different users' computing sites based on the user's role and the security requirements. We have presented our specification model using a detailed example of a document authoring activity.

Acknowledgements: This work was supported by NSF grant ITR 0082215 and EIA 9818338. The authors are with the Department of Computer Science, University of Minnesota, Minneapolis.

REFERENCES

1. E. Bertino, E. Ferrari, and V. Atluri. A Flexible Model Supporting the Specification and Enforcement of Role-based Authorizations in Workflow Management Systems. In *ACM Workshop on Role-based Access Control*, pages 1–12, 1997.

2. Mauricio Cortés and Prateek Mishra. DCWPL: A Programming Language for describing Collaborative Work. In *Proceedings of the ACM Conference on Computer Supported Cooperative Work*, pages 21 – 29, November 1996.

3. S.A. Demurjian, T.C. Ting, and B. Thuraisingham. User-Role Based Security for Collaborative Computing Environments. *Multimedia Review*, 4(2):40–47, Summer 1993.

4. Prasun Dewan and HongHai Shen. Access Control for Collaborative Environments. In *Proceedings of the ACM CSCW'92*, pages 51–58, 1992.

5. W. Keith Edwards. Policies and Roles in Collaborative Applications. In *Proceedings of CSCW'96*, pages 11–20, 1996.

6. Clarence A. Ellis, Simon J. Gibbs, and Gail Rein. Groupware: Some Issues and Experiences. *Communication of ACM*, 34(1):39 – 58, January 1991.

7. S. Foley and J. Jacob. Specifying Security for Computer Supported Collaborative Computing. *Journal of Computer Security*, 3(4):233–253, 1995.

8. Irene Greif and Sunil Sarin. Data Sharing in Group Work. *ACM Transactions on Information Systems*, 5(2):187–211, 1987.

9. R. Hayton, J. Bacon, and K. Moody. Access control in an open distributed environment. In *IEEE Symposium on Security and Privacy*, pages 3 –14, 1998.

10. Wei-Kuang Huang and Vijayalakshmi Atluri. SecureFlow: A Secure Web-enabled Workflow Management System. In *ACM Workshop on Role-based Access Control*, pages 83 – 94, 1999.

11. M.D.P. Leland, R.S. Fish, and R.E. Kraut. Collaborative Document Production using Quilt. In *Proceedings of CSCW'88*, pages 206–215, 1988.

12. Du Li and Richard Muntz. COCA: Collaborative Objects Coordination Architecture. In *Proceedings of CSCW'98*, pages 179–188, 1998.

13. Emil C. Lupu and Morris Sloman. Reconciling Role-Based Management and Role-Based Access Control. In *ACM workshop on Role-based Access Control*, pages 135–141, 1997.

14. Matunda Nyanchama and Sylvia Osborn. The Role Graph Model and Conflict of Interest. *ACM Transaction on Information System Security*, 2(1):3–33, February 1999.

15. P. Roberts and J-P. Verjus. Towards Autonomous Descriptions of Synchronization Modules. In *Proc. of IFIP Congress*, pages 981–986, 1977.

16. Ravi Sandhu, Edward Coyne, Hal Feinstein, and Charles Youman. Role-Based Access Control Models. *IEEE Computer*, 29(2):38–47, February 1996.

17. Ravi S. Sandhu. Transaction Control Expressions for Separation of Duties. In *Fourth Annual Computer Security Application Conference*, pages 282–286, December 1988.

18. R.T. Simon and M.E. Zurko. Separation of Duty in Role-based Environments. In *10th Computer Security Foundations Workshop*, pages 183 –194, 1997.

19. Suzanne Sluizer and Paul M. Cashman. XCP: An Experimental Tool for Managing Cooperative Activity. In *Proceedings of the 1985 ACM Thirteenth Annual Conference on Computer Science*, pages 251 – 258, 1985.

20. Roshan K. Thomas. Team-based Access Control (TMAC): A Primitive for applying Role-based Access Controls in Collaborative Environments. In *ACM Workshop on Role-based Access Control*, pages 13 – 19, 1997.

21. Jonathon E. Tidswell and Trent Jaeger. Integrated Constraints and Inheritance in DTAC. In *ACM Workshop on Role-based Access Control*, pages 93 – 102, July 2000.

22. Jonathan Trevor, Tom Rodden, and John Mariani. The Use of Adapters to support Cooperative Sharing. In *Proceedings of the Conference on Computer Supported Cooperative Work*, pages 219 – 230, 1994.

23. Anand Tripathi, Tanvir Ahmed, Richa Kumar, and Shremattie Jaman. Design of a Policy-Driven Middleware for Secure Distributed Collaboration. In *Proceedings of the International Conference on Distributed Computing Systems* , 2002.

24. N Yalelis, E Lupu, and M Sloman. Role-based security for distributed object systems. In *Proceedings of the 5th Workshops on Enabling Technologies: Infrastructure for Collaborative Enterprises*, pages 80–85. IEEE Computing Society, 1996.

Author(s) affiliation:

- **Anand Tripathi, Tanvir Ahmed, Richa Kumar, and Shremattie Jaman**

  Department of Computer Science & Engineering
  University of Minnesota
  Minnesota 55455, USA
  Email: [tripathi, tahmed, richa, jaman]@cs.umn.edu

Part II - Grid Coordination

6

The Use of Content-Based Routing to Support Events, Coordination and Topology-Aware Communication in Wide-Area Grid Environments

Craig A. Lee
B. Scott Michel

Abstract

The potential scale of a ubiquitous, internet-scale event service for coordination is staggering. While producers and consumers of events may be well-known to each other in many cases, there are many other situations where they may want to produce and consume events based on some type or properties through a *publish/subscribe* interface. Hence, *content-based routing* is a fundamental capability that will be necessary to support coordination in wide-area environments. Content-based routing also provides support for *interest management* in distributed simulations, and novel uses such as scoped, topology-aware communication semantics among the members of a group, including collective operations, filtering, compression, encryption, quality of service, and transcoding. Several implementation approaches are possible. These include (1) a traditional network of servers, (2) a middleware forwarding and routing layer in user-space, and (3) active networks. Content-based routing, however, which basically involves a form of associative matching, can be expensive to implement. While a good amount of work has been done in quantitatively evaluating a network of servers, the effectiveness of user-space middleware and active networks is just beginning to be investigated. Such work is clearly needed and motivated by the potential benefits of content-based routing in wide-area environments.

6.1 INTRODUCTION

Content-based routing, which basically involves a form of associative matching, is problematic. It provides fundamental and powerful capabilities with potentially far-reaching implications, but it is notoriously expensive to implement. As a world-wide compute infrastructure evolves to the point where it is genuinely ubiquitous, available, and easy to use, will the benefits of content-based routing find routine use? Or will they continue to be overshadowed by excessive compute times and excessive storage requirements? As grid computing evolves, however, to where virtually all resource types are routinely managed though a distributed information service, will users and applications be able to find a "sweet-spot" where the content-based routing space is constrained enough to be practical and provide a significant net benefit and capability that can't be accomplished any other way?

This paper concentrates on the first question, with the assumption that the third question will be answered in the positive. Given the complexity of large-scale associative matching, it is clear that problems sizes can always be made large enough to make content-based routing impractical. However, given the potential benefits of the expected, and novel, applications of content-based in the grid environment, it will become increasingly important to find out where the boundaries are.

Content-based routing can be supported in several different ways. As discussed in the next section, this includes existing implementations using (1) a network of servers, (2) a forwarding and routing layer that is below the applications but is still in user-space, and (3) active networks that are implemented at the system level. In wide-area environments, content-based routing could be used in an basic event service, for interest management in distributed simulations (a related field), and for managing communication topologies – an aspect of grid applications that will become increasingly important to manage latencies and how tightly coupled grid applications can be. These are discussed in Section 3. Besides the fundamental efficiency issue, there are a host of implementation issues concerning the use of emerging network technologies. Concepts such as minimum communication latency, multicast group construction and protocols, and overlays are all important topics. This is discussed in Section 4. Conclusions are given in Section 5.

Making large-scale grid applications and services (such as an event service) more network-efficient will be essential to their long-term success. In five to ten years time, the size and complexity of the deployed grid infrastructure and grid applications will be enormous by today's standards. Hence, it is imperative that these investigations are made today. We begin.

6.2 CONTENT-BASED ROUTING AND SUPPORTING TECHNOLOGIES

While the associative matching of content-based routing in the distributed environment of a grid carries significant complexity, it also provides fundamental capabilities that are possible no other way. Content-based routing can be implemented in several different ways. We begin by briefly discussing how content-based routing is done in existing publish/subscribe systems. We then discuss the use of *application-level routing and forwarding* and *active networks* to provide this capability.

6.2.1 Existing Publish/Subscribe Systems

Existing publish/subscribe services, such as event services, use a network of servers to accomplish filtering and matching. The CORBA Event Service [29], for instance, provides decoupled communication between producers and consumers using a hierarchy of clients and servers. Jini Distributed Events [26] define a set of classes that can be used to build an event service. Recent research projects in the area of scalable publish/subscribe services include Siena [5], Gryphon [1].

The fundamental problems faced by all of these systems can be partitioned into two parts: (1) the local matching problem, and (2) broker network design [11]. Efficient matching for all data on a single machine can be viewed as a database query problem where subscriptions are data and events are queries on subscriptions. This view gives rise to many approaches involving one-dimensional and multi-dimensional indexing schemes relying on, for example, hash tables, B-trees, and clustering. An interesting approach, the *pyramid technique* [2], converts an n-dimensional data space into $2n$ pyramids sharing the center point of the data space as their common apex. This provides a very useful mapping to B+-trees.

For a network of brokers or event servers, the problem comes down to the propagation of subscriptions, publication advertisements, and events throughout the network of participants. The common elements in these systems are subscription propagation up-stream to the publisher from the subscribers and publication advertisement propagation down-stream from the publisher to the subscribers. As one example, the *SCRIBE* [23] approach relies on the distributed B+-tree structure provided by the *Pastry* peer-to-peer infrastructure [8] to manage event propagation. *SCRIBE* publishers map event names and types onto fixed-length keys, typically SHA1-generated hash codes. Event subscriptions propagate up through the *Pastry* tree to the publisher, where the intermediate peers keep state with respect to the sibling tree branches to which the event data will be later replicated. In *SCRIBE*, *content-based filtering* or *routing* is an inherent infrastructure property and it is important to note that *Pastry* is reusable in other peer-to-peer application contexts. Other examples of content-based networking and proxy hierarchies are explored in [6] and [33].

101

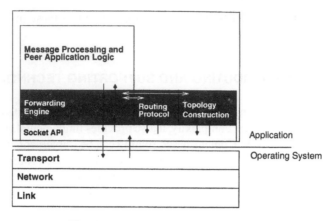

Fig. 6.1 *The FLAPPS forwarding layer.*

6.2.2 Application-Level Routing and Forwarding

A Forwarding Layer for Application-level Peer-to-Peer Services (*FLAPPS*) is a general-purpose peer-to-peer distributed computing and service infrastructure where content-based routing is achieved by interposing a routing and forwarding middleware layer between the application and the operating system, in user-space. This architecture, as shown in Figure 6.1, is comprised of three interdependent elements: peer network topology construction protocols, application-layer routing protocols and explicit request forwarding. *FLAPPS* is based on the store-and-forward networking model, where messages and requests are relayed hop-by-hop from a source peer through one or more transit peers en route to a remote peer. Hop-by-hop relaying relies on an interconnected application-layer network, in which some peers act as intelligent routers between connected peer groups and potentially between different peer networks as inter-service gateways. Routing protocols propagate reachability of remote peer resource and object names across different paths defined by the constructed peer network.

Resources and objects offered by a peer in a *FLAPPS*-based service are chosen from the service's name space; a name uniquely identifies a resource or object within the service. The service's name space is also assumed to be hierarchically decomposable so that collections of resources and objects can be expressed compactly in routing updates. A peer collects reachable names and name collections in a routing and forwarding table from the routing protocol updates. A request who's resource or object is not local to a peer is relayed by matching its resource's name to the longest prefix existing in the peer's routing and forwarding table and identifies an equivalent next-hop peer set. A next-hop peer set will contain at least one but possibly more than one peer, where each peer is considered an equivalent provider of the named resource or object. Multiple equivalent next-hop peer sets may exist for a given name, corresponding to different forwarding behaviors or alternate methods that choose a next-hop peer from the set. The default forwarding behavior, *first-of-m*, where m is

102

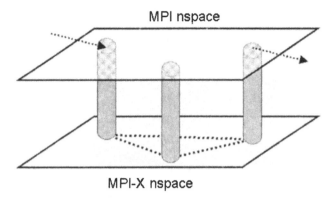

MPI nspace

MPI-X nspace

Fig. 6.2 *Routing through different services' peer networks.*

the cardinality of the next-hop peer set, always exists and relays a request to the first peer in the set. Other possible forwarding behaviors include *one-of-m* relaying, where the next hop peer is randomly chosen from the set, and *n-of-m* replication. Additional forwarding behaviors can be created from customization data included with routing protocol update. An example customization is *weighted one-of-m*, where a weighting factor is included with a name's reachability in the protocol's update and the next-hop peer is randomly chosen from the next-hop peer set in which the weights are normalized.

An important characterisitic of a *FLAPPS*-based service is transit peer transparency. *FLAPPS* does not assume that all peers are functionally homogeneous, but rather that transit peers are likely to be functionally heterogeneous. Transit peer transparency allows extending a service beyond its initial design as well as embedding specific functionality within the peer network. A request filtering peer in front of a group of source peers is a general example of embedded functionality; a file hoarding peer embedded in a file replication service is a more specific example. Transit peers may also act as gateways between services, extending the peer networks. Figure 6.2 illustrates two peer networks corresponding to a MPI service and an extended MPI-X service, linked via three gateway peers. The gateway peers would conceivably translate the event types published in MPI service to their MPI-X service equivalents before being relayed into the MPI-X service, or inhibit some event types originating in the MPI-X service from being relayed back into MPI service, or some combination thereof.

The *FLAPPS* infrastructure is designed to be flexible in a variety of ways. *FLAPPS* does not assume that one size fits all with respect to constructing a peer network so that multiple peer network topology construction protocols can be applied depending on their properties. An expanding multicast ring search is appropriate for discovering and building a peer network within a corporate or campus network, whereas a different mechanism is more applicable in wide-area networks. *FLAPPS* separates the style by which one of its routing protocol's propagates remote peer resource reachability and the customization data that affects next-hop peer selection. Routing protocol styles

103

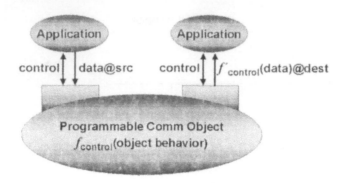

Fig. 6.3 *A programmable communication object.*

differ in the way that distance is measured, e.g. a distance-vector style protocol, which uses hop count, vs. a path-vector style protocol, which uses path length through routing areas. This avoids an explosion of reimplemented routing protocols while at the same time allowing customization. Routing protocol customizations provide the input to forwarding behaviors and subsequently influence how requests are relayed from one peer to another.

6.2.3 Active Networks

An active network is the concept of putting "intelligence", i.e., programmable capabilities, into network routers and switches [27]. Given the ability to manage state at the routers, network-centric functions are prime candidates for active networks. Such functions include firewalls and caching (e.g., web caching) [27], intrusion detection, dynamic rerouting (e.g., mobile hosts) [32], support of reliable multicast [14], fusion and fission of packets (filtering/merging and replication) and data transcoding [20]. Network management, to control operation and collect performance statistics, is another possible function. Since active networks provide a flexible infrastructure for network semantics, kernel-based QoS services (IntServ [4], DiffServ [3]) could also be supported.

In the fullest generality, active networks allow an application to associate a *control plane* with the flows in a *data plane* [16]. This enables the abstraction of a *programmable communication object* that is meaningful to the application, as illustrated in Figure 6.3. *Content-based routing* is a fundamental capability enabled by active networks that is managed in this manner. The routing of data is determined by application-defined fields in the payload data and is controlled by application signaling in the control plane which can change over the course of the computation.

This high-level notion can be mapped closer to a router architecture as illustrated in Figure 6.4. Here signaling control packets are propagated via some protocol. Each signaling packet may cause code to be executed at the router and may modify the filter state. This controls the (presumably higher) flow of data packets via filters that could be part of a kernel (such as Linux) or actually implemented in hardware. While

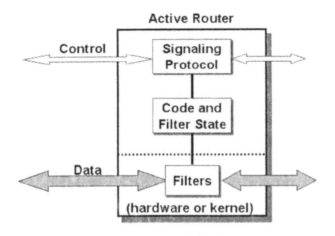

Fig. 6.4 *A generic active router.*

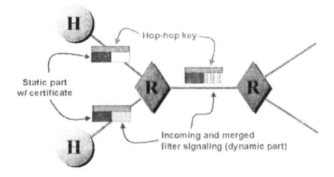

Fig. 6.5 *Signaling security between active router.*

this diagram shows separate control and data packet flows, in practice, these would be in-band together. A *packet classifier* would be used to elevate signaling packets to the appropriate level.

We note that this diagram bears a strong resemblance to the Integrated Services Node Architecture. Clearly, per-flow quality of service reservations require very similar router capabilities. This attests to the fundamental importance and wide applicability of making a data *and* control plane available to an application.

Security is a central issue for active networks. Typical network security models use a cryptological certificate to sign and verify packets. This requires that the packet does change between the source and ultimate destination. In active networks, however, their advantage lies in the fact that applications may require and specify how packets are to be modified en route. The signaling for some applications (such as interest management in distributed simulations) requires that the signaling payload be changed depending on what other signaling packets a router has seen. Figure 6.5 illustrates

105

how a certificate may be used to protect the static part of a signaling packet, e.g., source and destination addresses and ports, while a key is used to provide *hop-by-hop integrity checking* on the dynamic part.

Performance is another important issue for active networks. In addition to packet classification, an active router may have to parse the packet payload, execute code, and maintain state over a memory bus. To meet such compute requirements, active routers will be architected with network processors such as the IXP1200 [25]. In the core of the network, however, backbone routers will continue to be dedicated to servicing high-volume aggregate flows. Hence, like QoS services, active network services will be deployed at the periphery of networks.

6.3 THE USE OF CONTENT-BASED ROUTING

Content-based routing is clearly an important, fundamental capability. This importance is underlined by the fact that it will support a variety of functions in wide-area environments. It is clear that a grid event service is needed and that content-based routing is directly applicable. We discuss this next. We subsequently discuss how content-based routing can be used for the closely related task of *interest management* in distributed simulations and also for *topology-enabled* communication.

6.3.1 A Grid Event Service

Events are a fundamental building block in many applications and systems. While events can be viewed in different ways, they can be operationally viewed as "little messages". The properties and capabilities of events and event delivery, however, have serious implications for both implementation and use. Simply the fact that events are typically delivered asynchronously to an application means that they can be used to implement and enforce a wide variety of basic application and system behaviors.

Examples of these behaviors include selective dissemination of information (e.g., stock quotes), notification of service availability or job completion, on-line auctions, expiration of time periods (e.g., leases), and system monitoring and management. In grid terms, events will be used for performance monitoring. The Grid Monitoring Service Architecture [30] defines an event service for performance events. Grid application frameworks, such as Cactus, will support specialized "thorns" for notification, e.g., process completion (or other significant event) will communicate with an SMS Gateway (Short Message Service) send a small amount of text to a cellphone. Collaborative environments depend heavily on events. Physical events, such as keystrokes or navigational movements, must be propagated among all participants to maintain a consistent virtual world. Discrete simulations deal explicitly with events to propagate simulated time. Finally, any type of grid component architecture based on remote procedure call (RPC) or remote method invocation (RMI) will require an event service to achieve some level of fault tolerance [17]. Besides "routine" asynchronous events denoting desired events, e.g., job completion, events can be used to alert failure of networks, hosts or processes.

6.3.1.1 Event Representation and Discovery

In any event service, a well-known representation schema must be defined. While it is possible to define some base class of events, applications will need some extensible scheme whereby new types of events can be defined. As an example, [9] describes an XML-based schema for performance events based on simple attribute-value pairs, while [24] describes an extensible "measurement container" with metadata. Such an approach could be generalized to any type of event.

We should note that while events can be considered "small messages", such extensibility means that events could have unbounded size. Below some size threshold, events could enjoy certain properties, e.g., lower latency by fitting into one UDP packet. If an event can carry arbitrary data and metadata, applications should be smart enough to decide (1) if a large data volume is necessary, or (2) if a reference to the data can be attached such that the event consumer can decide if it is worth it to actually access the data.

Events have producers and consumers. In the simplest case, an event has one producer and one consumer and they are well-known to each other. In many cases, however, producers and consumers do not initially know each other. Consumers may also be interested in any events of a certain type or producers of a certain type. In this case, there must be a *registry and discovery service* whereby producers and consumers can register what they produce or consume and discover relevant partners. Naturally this requires a metadata schema for producers, consumers, and events. Registries could be local or hierarchical to manage scalability in addition to caching producer/consumer handles based on actual use. We note that while producers and consumers could *push* their identities into a registry service for subsequent discovery, a *discovery agent* could also probe the grid environment to find producers/consumers and *pull* their identities into a registry (much like web search engines discover and catalog web content).

6.3.1.2 Event Delivery

Besides the basic schemas for representing events and managing their use, events have to be transported between producers and consumers. Event delivery is asynchronous and, in the base case, does not require a reply. Beyond this, single event delivery semantics can be classified as deliver at most once (unreliable delivery), exactly once (reliable delivery), any number of times (idempotent). (Clearly this has direct implications for the fault models that can be supported: in the presence of network or host failure, can an event be delivered more than once?) Messages are intended to be lightweight with low latency. While latency can be improved by relying on different underlying network mechanisms (e.g., UDP vs. TCP) it can also be addressed by using priorities and service classes. An actual lower bound to the latency would require network QoS. A *time-to-live* can be used to enforce an upper bound on event life. In this case, even undelivered events can be discarded at expiration.

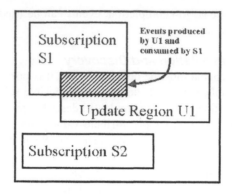

Fig. 6.6 *Hyperbox intersection in two dimensions.*

While many events will only have one producer/consumer and only require point-to-point communication, many useful coordination semantics require an *event service.* An event service between the producers and consumers can be used to buffer, filter, modify, merge, replicate and multiplex events based on policy. Such semantics can be part of named *event channels* that are instantiated by applications. Events that are replicated have multiple consumers. Events that are merged can be thought of as having multiple producers. With the use of named event channels, producers and consumers could interact with each other via the channel without explicit knowing each other's identity. What becomes important is the name of the channel and the event types it handles.

Event channels can also be used to support *associative addressing* or a *publish/subscribe* service. Producers publish events with attached attributes and consumers subscribe to certain attribute types. Events can propagate without producers knowing where they go, or consumers knowing where they came from. This naturally requires associative matching and is functionally similar to Linda tuple spaces and JavaSpaces. In the case of a wide-area, distributed matching engine, content-based routing becomes directly applicable.

6.3.2 Distributed Simulations

The notion of a general grid event service is strongly related to events in a large-scale, distributed simulation. In a general event service, events may be time-stamped but an event's primary importance is that it arrives and is processed as soon as possible. The time-stamp carried on the event may be secondary. In a discrete-event simulation, events carry a stamp of *simulated* time that determines how events are processed, the progression of simulated time and, in turn, the progression of the entire computation.

Distributed simulations can also use a publish/subscribe service to manage the communication of events. The High-Level Architecture (HLA) [28] defined by the Defense Modeling and Simulation Office (DMSO) provides such a service officially called *Data Distribution Management* but is also known as simply *interest manage-*

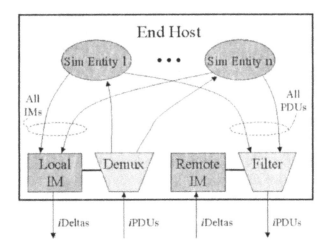

Fig. 6.7   *Endhost handling of Interest Maps and Program Data Units.*

ment. Interest management in the DMSO HLA is defined on the model of *hyperbox intersection.* A federated simulation defines a set of event attributes. Simulated entities can produce events in an *update region* and express interest in a *subscription region* defined by subranges across the set of events attributes. As illustrated by Figure 6.6 in two dimensions, only those events that fall within a hyperbox intersection must be delivered from the producer to the consumer. Of course, as the event dimensionality and the number of simulated entities increases, the number of unique intersections grows combinatorially. This situation can be alleviated by aggregating smaller intersections into fewer larger one, thus reducing the number of intersections at the cost of allowing a few events to be delivered to end-hosts that do not need them.

In a distributed environment, this type of publish/subscribe-based interest management can be cast as a content-based routing problem. This can even be simplified to a filtering problem since for each interface on a router, the router most only decide to forward the packet or not. Figure 6.7 illustrates how an endhost handles Interest Maps (IMs) and Program Data Units (PDUs) (to use the HLA terminology). Simulated entities publish their interest regions which are combined into a local IM for the host. Interest changes, or deltas, are published to the communication service (e.g., the network). Deltas from all endhosts propagate through the network and are combined as necessary. Eventually an interest delta may get delivered to an endhost's Remote IM. Whenever a simulated entities produces an event (a PDU), it is filtered against the Remote IM. If the PDU passes, this means that at least one other endhost needs this event, but the sending host does not know who it is. Similarly, PDUs will occasionally be delivered to the endhost as a result of previously published interest deltas. The receiving endhost may not know where the PDU came from, but it is demuxed and delivered to the appropriate simulated entity.

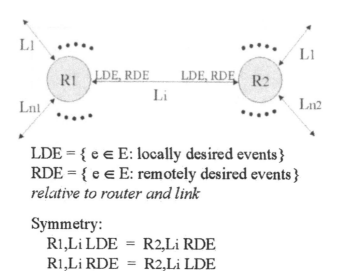

LDE = { e ∈ E: locally desired events}
RDE = { e ∈ E: remotely desired events}
relative to router and link

Symmetry:
   R1,Li LDE = R2,Li RDE
   R1,Li RDE = R2,Li LDE

Fig. 6.8 *Router handling of Interest.*

Figure 6.8 illustrates how interest deltas can propagate through the network. A router maintains a set of "locally desired" and "remotely desired" events relative to each link. Here, "locally desired" means that the router wants to receive events of type e from that link, while "remotely desired" means that the router must forward events of type e down that link. Note that for two adjacent routers, the local and remote desired event sets are the same. These sets change as interest deltas propagate through the network.

This type of interest management, i.e., content-based routing, has been successfully demonstrated using an active network testbed as part of the DARPA Active Networks program. Using Linux hosts as active routers (much like that shown in Figure 6.4), interest management was implemented using the Linux packet filtering capability [35]. Filter signaling was integrated in to the Georgia Tech HLA/RTI such that a ModSAF simulation running on top could transparently use the active network filtering capability [34]. Figure 6.9 illustrates the engagement between two tanks and a jet fighter. A terrain map showing the engagement is in the upper-left. For each of the simulated entities on the terrain map, their corresponding update and subscribe regions are shown in the diagram in the upper-right. The three strip charts show the number of data packets delivered to each simulated entity. The packets are color-coded to identify whether they were wanted and received (blue), neither wanted nor received (green), received but not wanted (yellow), and wanted but not received (red). While most packets are colored green, as the interest boxes for the Red Tank Platoon A and the Blue Airstrike intersect, packets colored blue start to flow between these hosts.

Active networks can also be used in other areas of distributed simulation, e.g., *Time Management* [19]. Time management is essentially an instance of the Distributed

110

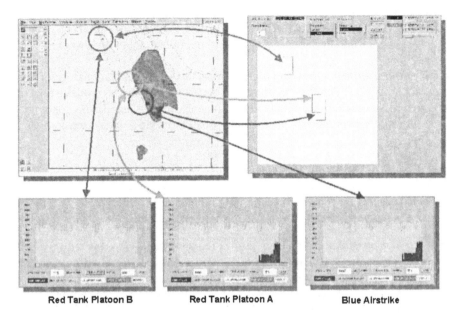

Red Tank Platoon B   **Red Tank Platoon A**   **Blue Airstrike**

Fig. 6.9 *Tank and Fighter Jet engagement. (From [34].)*

Termination Detection problem [21] but with an associated reduction that must include all endhost and in-flight messages.

6.3.3 Topology-Enhanced Communication

For networked applications, understanding and utilizing network topology will be increasingly important since overall grid performance will be increasingly dominated by propagation delays. That is to say, in the next five to ten years and beyond, network "pipes" will be getting fatter (as bandwidths increase) but not commensurately shorter (due to latency limitations) [18]. Early experience with MPICH-G2 for distributed, message-passing applications indicate that topology-awareness is crucial for efficient collective operations [13]. Rather than being governed by $O(n \log n)$ messages across the diameter of the network, as is typical, topology-aware collective operations could be governed by just the average diameter of the network.

Such topology awareness can be extended to include that of shared-memory nodes, machines, and clusters.

Topology-awareness also means that applications would be able to manage their communication based on fundamentally different capabilities such as *content-based* or *policy-based* routing. A traditional multicast group, for example, builds relevant routing information driven by the physical network topology. Content-based routing would enable an application to control the communication scheduling, routing and filtering based on the application's dynamic communication requirements within a given multicast group, rather than always having to use point-to-point communication. We

111

note that an application can be topology-enabled without being explicitly topology-aware. Being topology-enabled does imply, however, that an application can control some lower layer of middleware or services that are, in fact, topology-aware.

6.3.3.1 Possibilities How can high-performance grid applications exploit the network topology on which they are running? Topology-awareness enables a host of capabilities; some of which are simply more efficiently implemented using topology-awareness, while others are not possible any other way [16]. We discuss several possibilities.

- *Enhanced network-centric communication.* Many network-centric (rather than application-centric) functions could benefit from topology-awareness. These include caching (web caching), filtering, compression, encryption, quality of service, data-transcoding, or other user-defined functions.

- *Collective operations.* Applications may require synchronous operations, such as barriers, scans and reductions. These operations are typically implemented with a communication topology based on point-to-point operations. For performance in a wide-area network, it is crucial to match these operations to the topology defined by the physical or virtual network.

- *Communication scope.* Collective operations also imply a *scope* for communication. A scope can be associated with a *named topology* such that multiple scopes can be managed simultaneously for separate computations.

- *Content-based and Policy-based Routing.* As discussed under active networks, content-based routing allows an application to determine routing based on application-defined fields in the data payload. Policy-based routing is also possible. Examples of this include routing to meet QoS requirements and message consistency models where a policy must be enforced on the message arrival order across some set of end-hosts.

- *Tuple spaces.* A tuple space defines a communication scope with *associative matching* on the type and number of *put* and *get* arguments. While associative matching for an application-level problem in the network would be excessive, a distributed matching engine could use content-based routing to deliver tuples to where matches are more likely to be achieved.

We now discuss a particular programming tool and possible enhancements to cope with the grid environment.

6.3.3.2 An Example: Topology-Enabled MPI MPI is a widely used message-passing tool. As mentioned above, topology-awareness is critical to maximizing performance in a large-scale, networked environment. If this can be done such that minimal or no changes are required at the end-user level, i.e., with a low barrier to acceptance, then a topology-enabled MPI could easily find a large user base. Some work has already been done in this regard. The MagPIe system [15] transparently accommodates wide-area clusters by minimizing the data traffic for collective operations over the slow links. A similar concept can be implemented using application-level or active networks routing but with greater ramifications.

MPI uses the concept of a *communicator* to scope communication and collective operations. The key concept is this: *a communicator can be managed using content-based routing within a multicast session.* We will call this an *active communicator.* In much the same way that content-based routing is used for interest management, communicator routing can be done based on the communicator *rank* or *index* and the type of MPI operation. Topology information is built as part of the multicast group using established techniques such that a communicator essentially defines a virtual network. Of course, initial construction of the communicator group is a cost that must be amortized over the life of the communicator. Once constructed, however, end-hosts need no other information to be topology-enabled; they simply write messages to a single active multicast group which handles routing. We note that in five or ten years, very large MPI codes could be deployed over thousands of processors, i.e., communicators could be sized in the thousands. In this case, active network communicators would have the additional benefit that an operating system would not be risking file descriptor depletion when supporting such large codes.

The unique advantage of an active communicator is that the multicast topology can be used for collective operations. Rather than implementing a barrier using *O(logn)* message stages that chain the wide area latencies (as MPICH does), an active communicator can achieve synchronization in the network. A number of algorithms for this are possible that follow those in the area of distributed termination detection [19]. One such algorithm uses a distinguished root node (active router) in the network. In this case, barrier and reduction latencies are determined by a single traversal of the diameter of the active communicator. (We note that while reduction operators may be associative in theory, they may not be in practice due to rounding or over/underflow conditions. To be wide-area optimal, however, such concerns would have to be relaxed.) MPI broadcast operations would certainly benefit from the use of multicast, as would other one-to-all or all-to-all operations.

Communicators are, of course, not static. *Inter*-communicators and *intra*-communicators can be dynamically created, duplicated, or split. These operations would initiate signaling in the active network to create the appropriate routing. As mentioned above, initialization signaling for a communicator must be amortized over its life.

6.4 IMPLEMENTATION ISSUES

Even within the three basic supporting technologies discussed in Section 2 (network of servers, user-level middleware, and active networks), there are a host of implementation issues that will directly affect the behavior and performance of application uses of content-based routing. A basic requirement is that communication latencies should be minimized. As mentioned earlier, wide-area computations and services will become increasingly dominated by simple propagation delays. Such delays will limit the discoverability of information (events) and resources, and the tightness of interaction. Nonetheless, many techniques can be used to tolerate latencies, including that of topology construction.

In the most general case, minimizing the communication delays among a network of server or routers is an instance of the Steiner Tree Problem [10]. Given a graph with weighted edges, one must find a minimum weighted subgraph connecting a subset of the vertices call *terminals*. This problem has many variants and applications such as VLSI routing, wirelength estimation, and even phylogenetic tree reconstruction in biology. While the Steiner Tree Problem is *NP*-hard, a number of approximations exist in polynomial time to within ≈ 1.55 of optimal [22].

In a more practical sense, topology construct will depend on current techniques for multicast group construction and routing. PIM (Protocol Independent Multicast) [7] is widely used. In *sparse mode*, a single routing tree is constructed to connect members of a multicast group. In *dense mode*, there is a tree per source, i.e., a separate routing tree for every member of the group. While dense mode can achieve a lower average latency among all members (by using more direct routes between any pair of members), sparse mode requires less multicast routing state to be maintained (which is especially important for large-scale, wide-area deployment). In sparse mode, the primary issue is how to build the tree such that latency is minimized, or available bandwidth is maximized (since traffic between all member pairs is concentrated on the tree edges). In dense mode, content-based routing would require even more state to be carried as subscription and publication advertisements would have to be propagated up and down all trees.

For any implementation approaches, it is certain that content-based routing will not be deployed everywhere at once; rather isolated servers or routers will be available at a few sites interested in using the capability. For this reason, *network overlays* provide a useful approach for making such isolated resources look like part of a more unified whole. By using IP tunnels, such resources could have virtual direct connections through the existing IP fabric. Systems for managing overlays, such as the X-Bone [31], could also be used to support application-specific scoped communication, such as for an active MPI communicator. As always, the topology of such overlays will be important. Topologies such as rings or stars (as in the X-Bone) are useful, but overlays could also be self-organizing [12].

6.5 CONCLUSIONS

We have investigated the implications and issue of using content-based routing in wide-area environments. Content-based routing – and by implication, associative matching – is a fundamental function that can provide capabilities that are possible no other way. Content-based routing can be used to support publish/subscribe semantics for event models; for both simple events (such as system events) and also time-stamped events (such as for distributed simulations). It is also possible to use content-based routing for novel applications, such as "scoped communication". This is meant to denote a "communication object" that can be instantiated by an application and provide specific semantics for the members of the object. Such semantics could be topology-aware and include a publish/subscribe service (e.g., content-based routing), collective operations, filtering, compression, encryption, quality of service, transcoding, etc. Such a communication object could, in fact, be associated with an indexed communication space, i.e., an MPI communicator.

The challenge, of course, is that content-based routing is difficult to implement and use efficient in a large-scale. Nonetheless, different implementation and deployment techniques may allow the size of specific routing problems to be constrained to be practical and beneficial. Possible implementation approaches include (1) a traditional network of servers, (2) a middleware forwarding and routing layer in user-space, and (3) active networks. Each of these approaches offer ways to limit the overhead and potentially provide a significant benefit. While a good amount of work has been done in quantitatively evaluating a network of servers, the effectiveness of user-space middleware and active networks is just beginning to be investigated. Much work remains to be done in developing tools such as analytical models and proof-of-concept prototypes. A prototype implementation would force the consideration of practical issues for real-world deployment, such as unreliable delivery, dynamic routing, packet delivery order and, of course, security.

Deploying such capabilities in a wide-area environment will certainly be challenging. Nonetheless, making applications as network-efficient as possible is necessary. This can be done by not only providing enhanced services, such as content-based routing, but also by enabling applications to be as tightly coupled as possible, i.e., by making them topology-aware. While the fundamental limitation of speed-of-light propagation delays will always be present, these techniques can ameliorate the problem and move the boundary of acceptable performance. Ultimately, these technologies could be coupled with asynchronous programming and execution models to extract the best possible performance from large-scale networked applications.

REFERENCES

1. G. Banavar et al. An efficient multicast protocol for content-based publish-subscribe systems. In *IEEE International Conference on Distributed Computing Systems*, 1999.

2. S. Berchtold, C. Böhm, and H.-P. Kriegel. The pyramid technique: Towards breaking the curse of dimensionality. In *SIGMOD 98*, 1998.

3. S. Blake et al. *An Architecture for Differentiated Services*. IETF RFC 2475, 1998.

4. R. Braden, D. Clark, and S. Shenker. *Integrated Services in the Internet Architecture: An Overview*. IETF RFC 1633, 1994.

5. A. Carzaniga, D. S. Rosenblum, and A. L. Wolf. Achieving scalability and expressiveness in an internet-scale event notification service. In *Symposium on Principles of Distributed Computing*, pages 219–227, 2000.

6. A. Carzaniga and A. L. Wolf. Content-based networking: A new communication infrastructure. In *NSF Workshop on an Infrastructure for Mobile and Wireless Systems*, Scottsdale, AZ, Oct. 2001.

7. S. Deering et al. The PIM Architecture for Wide-Area Multicast Routing. *IEEE/ACM Transactions on Networking*, 4(2):153–162, April 1996.

8. P. Druschel and A. Rowstron. PAST: A large-scale, persistent peer-to-peer storage utility. In *HotOS VIII*, May 2001.

9. D. Gunter and W. Smith. Schemas for grid performance events. *http://www-didc.lbl.gov/GGF-PERF/GMA-WG/papers/GWD-Perf-1-1.pdf*, October 2000.

10. F. K. Hwang, P. Winter, and D. S. Richards. *The Steiner Tree Problem*. Elsevier Science, 1992. ISBN 044489098X.

11. H.-A. Jacobsen and F. Llirbat. Publish/subscribe systems. In *17th International Conference on Data Engineering*, 2001. Tutorial.

12. S. Jain, R. Mahajan, D. Wetherall, and G. Borriello. Scalable Self-Organizing Overlays. *http://www. cs.washington.edu/homes/sushjain/quals.ps*, 2001.

13. N. Karonis. Personal communication, November 6, 2000.

14. S. Kasera et al. Scalable fair reliable multicast using active services. *IEEE Network Magazine*, January/Febuary 2000. Special issue on multicast.

15. T. Kielmann et al. MagPIe: MPI's collective communication operations for clustered wide area systems. In *Symposium on Principles and Practice of Parallel Programming*, pages 131–140, May 1999. Atlanta, GA.

16. C. Lee. On active grid middleware. *Second Workshop on Active Middleware Services*, August 1, 2000.

17. C. Lee. Grid RPC, Events and Messaging. *http://www.eece.unm.edu/~apm/WhitePapers/APM_Grid_RPC _0901.pdf*, September 2001.

18. C. Lee and J. Stepanek. On future global grid communication performance. *10th IEEE Heterogeneous Computing Workshop*, May 2001.

19. C. Lee, J. Stepanek, C. Raghavendra, and K. Bellman. Time management in active networks. In *3rd International Workshop on Active Middleware Services*, August 6, 2001.

20. U. Legedza, D. Wetherall, and J. Guttag. Improving the performance of distributed applications using active networks. *IEEE INFOCOM*, April 1998.

21. J. Matocha and T. Camp. A taxonomy of distributed termination detection algorithms. *J. of Systems and Software*, 43(3):207–221, 1998.

22. B. Robins and A. Zelikovsky. Improved Steiner Tree Approximation in Graphs. In *SIAM-ACM Symposium on Discrete Algorithms (SODA)*, pages 770–779, January 2000.

23. A. I. T. Rowstron, A.-M. Kermarrec, M. Castro, and P. Druschel. SCRIBE: The design of a large-scale event notification infrastructure. In *Networked Group Communication*, pages 30–43, 2001.

24. W. Smith and D. Gunter. Grid information service schema for grid events. *http://www-didc.lbl.gov/GGF-PERF/GMA-WG/papers/GWD-Perf-2-1.pdf*, October 2000.

25. T. Spalink, S. Karlin, and L. Peterson. Evaluating network processors in IP forwarding. Technical Report TR-626-00, Princeton University, 2000.

26. Sun Microsystems, Inc. Jini EV – Distributed Events. *http://www.sun.com/jini/specs/jini1.1html/event-spec.html*, 2000.

27. D. Tennenhouse, J. Smith, W. Sincoskie, D. Wetherall, and G. Minden. A survey of active network research. *IEEE Communications Magazine*, 35(1):80–86, January 1997.

28. The Defense Modeling and Simulation Office. The High Level Architecture. *http://hla.dmso.mil/*, 2000.

29. The Object Management Group. CORBA 3 Release Information. *http://www.omg.org/technology/corba/ corba3releaseinfo.htm*, 2000.

30. B. Tierney et al. Grid monitoring service architecture. *http://www-didc.lbl.gov/GGF-PERF/GMA-WG/papers/ GWD-Perf-6-1.pdf*, February 2001.

31. J. Touch. Dynamic Internet Overlay Deployment and Management Using the X-Bone. *Computer Networks*, 2001.

32. D. Wetherall, J. Guttag, and D. Tennenhouse. ANTS: A toolkit for building and dynamically deploying network protocols. *IEEE Openarch '98*, April 1998.

33. H. Yu, D. Estrin, and R. Govindan. A hierarchical proxy architecture for internet-scale event services. In *WETICE '99*, June 1999.

34. S. Zabele, B. Braden, S. Murphy, and C. Lee. ANETS distributed simulation. Briefing for DARPA Active Networks demonstration. Available from *http://www.dsic-web.net/ito/meetings/anets2000dec/index.html*, December 7, 2000.

35. S. Zabele and T. Stanzione. Interest management using an active network approach. In *Spring Simulation Interoperability Workshop*, March 26-31, 2000.

Author(s) affiliation:

- **Craig A. Lee**

  Computer Systems Research Department
  The Aerospace Corporation
  California 90245, USA
  Email: lee@aero.org

- **B. Scott Michel**

  Computer Science Department
  University of California, Los Angeles, USA
  Email: scottm@cs.ucla.edu

7

The Complexity of Scheduling and Coordination on Computational Grids

Dan C. Marinescu
Gabriela M. Marinescu
Yongchang Ji

Abstract

In this paper we develop a model for scheduling and coordination on a computational grid and study their complexity based upon the concept of Kolmogorov complexity. The model supports a qualitative analysis of different aspects of scheduling and coordination and allows us to draw conclusions regarding the effect of crossing the boundaries of autonomous domains, the role played by the granularity of resources and the task dependencies.

7.1 INTRODUCTION

As human-made systems become more complex it is increasingly more important to provide some quantitative measures of the complexity of an object. We expect such measures to be consistent with our intuitive notions of how difficult is to understand the behavior of the system and how difficult it is to control its functions, to optimize its behavior, to rank systems with similar functions and characteristics, and eventually assess if a system posses some desirable properties.

119

The term complexity is used in this paper in a slightly different sense than the one favored by the theoretical computer science; it relates to the description of an object and to the Kolmogorov complexity. Let us turn our attention to the description of an object. Such a description is expected to consist of structural, functional, and possibly other important properties of the object. We have reasons to believe that the structure of an object determines its functions. This is a well known principle in natural sciences; for example, to understand the functions of a biological material one needs first to discover and understand its structure.

One possible quantitative measure of the complexity of an object is the length of its description. The more bits we need to describe the structure, the more likely is that the object is more complex.

To reduce the complexity of an object we need to compress its description, a phenomena exploited by nature. As we all know, biological materials carry with them the genetic blueprint; the larger the structure, the more intricate is this description and the more difficult it is to pack the genetic description in the limited volume of a protein molecule. Yet, the *genetic economy* allows the nature to build very complex biological systems and limit the size of the genetic blueprint. To illustrate this phenomena let us turn our attention to structural biology and consider very large macromolecules like viruses. A virus is a complex structure with millions of atoms and hundreds of thousands of aminoacids. Most viruses enjoy some degree of symmetry, they are built by repeating an *asymmetric unit* throughout the entire structure. A spherical virus has an icosahedral symmetry; there are 60 symmetry operators and an icosahedron can be built by applying these transformation to the asymmetric unit. Thus the genetic economy helps reduce the size of the genetic blueprint that now has to describe only the asymmetric unit.

The lesson we should learn from nature is that to build increasingly more complex objects we need a few types of building blocks and a limited set of composition operators. A system built following these rules is likely to be much more scalable than one where the component space is unstructured.

Building a computational grid seems a straightforward application of the principles outlined above; we take a small number of classes of computer systems and link them together to form the larger structure, a grid. Unfortunately, the problem is slightly more complicated because the *bare grid* resulting by interconnecting computers with each other is of little use as such; a computer becomes interesting only when we run some applications on it. The object we are really interested in is the ensemble consisting of grid plus applications. Each application shapes the resource rich environment provided by the bare grid into an *application grid*. Scheduling and coordination are two important mechanisms used to transform the bare grid into an application grid. Now we are dealing with a very complex object because the application has to relate its internal structure and needs with the resources that belong to different entities.

We are concerned with complex applications consisting of multiple dependent tasks whose relationships with each other are described by an activity graph. The graph contains also information about the resource requirements of each task. The scheduling problem is further complicated by the fact that the resource requirements are known only approximately; the actual execution time of a task, the amount of data transferred over the network during the execution of the task, or its explicit I/O requirements may differ from the ones specified by the activity graph.

In this paper we propose a model to study the complexity of scheduling applications consisting of dependent tasks on a computational grid. Even though the model is too sophisticated for a quantitative analysis it is well suited for a qualitative analysis. The model is consistent with our intuition that the finer the granularity of the resources allocated to each task, the more complex the tasking structure of the application and the larger the number of autonomous domains involved, the more complex the scheduling problem becomes.

The paper is organized as follows. Section 7.2 presents a historic view of our scheduling and coordination leading to the requirements posed by execution of complex computational tasks on grids. In Section 7.3 we discuss Kolmogorov complexity and derive an upper bound for the Kolmogorov complexity of a stationary Markov chain. A model for coordination and scheduling on a computational grid is introduced in Section 7.4.

7.2 RESOURCE SHARING ACROSS AUTONOMOUS DOMAINS AND COORDINATION OF DEPENDENT TASKS

Scheduling and coordination are among the more difficult problems encountered by high performance distributed computing on clusters and computational grids. High performance distributed computing refers to compute- and/or data-intensive applications whose needs exceed by several orders of magnitude the resources available on a single system.

Clusters are heterogeneous collections of commodity computers connected by high speed networks; typically, a cluster is under the control of a single administrative authority. *Computational grids* are large-scale collections of heterogeneous systems located within distinct autonomous domains and interconnected by the Internet.

Here the term *autonomous domain* refers to a collection of computers and communication systems under the control of a single authority enforcing security, networking, and access policies and standards. The term *resource* is used in a wide sense, it means hardware and software resources, services, and content. *Content* generally means some form of static or dynamic data or knowledge. *Autonomy* implies that resource sharing requires cooperation between the administrative authorities of each domain [5].

Scheduling computing resources is a process of matching needs with availability; a computation needs resources and a system or a group of systems make these resources available to the computation for variable periods of time. Scheduling is a loaded concept, it covers both policies and the means to enforce them. Moreover, scheduling requires some agreement between the group of producers and the consumers and the means to control the behavior of non-cooperative entities in each group.

Coordination is the dual of scheduling; it means supervision of complex activities, making sure that dependencies among the individual tasks are observed, that a course of action leading to optimal performance is taken, that data security and confidentiality is preserved, and that the final objective is achieved even when individual activities fail. Coordination and scheduling are strongly correlated. Loosely speaking, coordination describes what needs to be done, while scheduling determines the exact time and place where each activity may take place.

Sharing computer resources is driven by economics and has evolved in time, see Figure 7.1. In early days, computer systems were relatively expensive and one system was usually shared among the members of the same organization; sharing policies were established by a single administrative authority. An application was monolithic, its tasking structure was transparent to the scheduler.

With the introduction of commodity computing, a one-to-one mapping of users to systems became the norm and the more difficult problems posed by computing resource sharing were partially avoided by the massive introduction of personal computers (PCs).

Nowadays, we wish to combine the two paradigms and build an infrastructure allowing a virtually unlimited number of users and applications to share a large numbers of autonomous systems in a transparent manner, subject to some level of quality of service guarantees. Moreover, we wish to carry out complex computational tasks with strong dependencies among activities and a wide spectrum of resources needed for each activity.

The many-to-one scheduling model with independent tasks dominated the early days of computing; the model is still used in high performance computing on one of a kind systems, the so called supercomputers. In the original version of this model the scheduling was restricted to the CPU, an entire application was mapped into one processe regardless of its tasking structure, the processes were assumed to be independent entities and the resources were allocated to processes; in the general case no timing constraints were observed.

In the many-to-one scheduling model the system scheduler mediates among competing processes by allocating each process a time slice. There are no entities capable to negotiate resource allocation on behalf of an application.

No significant advances in schedulers were required for the one-to-one model because we are still confined to a single autonomous domain. The competition for system resources is less intense, we can identify a dominant activity and allocate resources preferentially to it.

The one-to-one scheduling model coexists with a version of the many-to-many scheduling model tailored to support client-server execution. In this case some weak task dependencies may exist; for example, prior to contacting a Web server, a browser may need to map hosts names to IP addresses, a service supported by DNS (Domain Name Services). We call this a *weak dependency* because: (a) the resource requirements are well defined and the resource allocation policies at the server site are under the control of a single entity or of cooperating entities, as is the case of clusters of servers discussed below. (b) services typically do not interact with each other. For example, before sending an HTTP request to a Web server the client may need to map the host name to an IP address, a service provided by DNS (Domain Name Services); the two services are strictly sequential since the HTTP request is sent after the response to the DNS query has been received. Moreover, the client may have

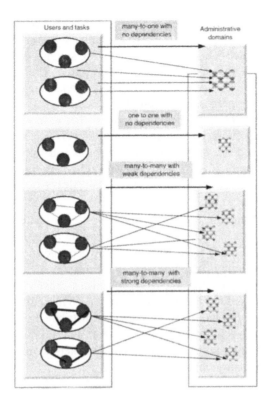

Fig. 7.1 Mappings of consumers of resources and administrative domains. The terms strong and weak dependencies refer to both the inter task dependencies and to the fact that an individual task may or may not require cooperation of multiple autonomous domains.

already cached the mapping of host name to the IP address and there is no need for DNS access.

Additional complexity is required at the client side and now we need to have an application coordinator. Of course, the coordinator may be a script or simply a program issuing Remote Procedure Calls (RPCs), but performance and reliability considerations plead for more intricate coordination mechanisms.

Often, this version of the many-to-many scheduling model is extended to support scalability, priority scheduling, and some form of QoS. To ensure scalability one replaces a single server with a cluster of servers; for priority scheduling we need a classification mechanism working in concert with a multi-queue scheduler; quality of service guarantees can be ensured by an admission control policy implemented by a front end of the cluster.

Yet, none of these extensions increases significantly the conceptual difficulties of either scheduling or coordination. Individual requests are processed by independent threads of control. Each request may be queued for all locally shared resources such as CPU, communication network interfaces, secondary storage. At each queue a request

123

should be processed according to its priority, a proposition difficult to implement with the current generation of operating systems which do not preserve priorities once a process crosses the kernel boundary. Most operating systems process requests for system services on a FIFO basis ignoring the priority of process issuing the system call. The queuing models of such systems involve networks of queues, one queue for each resource needed for processing a request.

Let us now examine the last scheduling model. The term *strong dependencies* refers to the fact that execution of each task requires resources in several autonomous domains, there are precedence relationships among tasks, and tasks may communicate with each other at execution time. This model can be further refined to isolate a very difficult case, the so called co-scheduling of tasks, but this is beyond the scope of this paper.

We believe that a quantum leap in complexity of coordination and scheduling heuristics and algorithms is necessary to cross from the many-to-many with weak dependencies models to many-to-many with strong dependencies models. Coordination becomes considerably more complex; performance, security, and reliability considerations require more sophisticated mechanisms capable of: (a) supporting resource discovery; (b) enforcing task dependencies, (c) managing data distribution and gathering of partial results, (d) enforcing some distributed security policies, (e) detecting the failure of an activity and replacing it with with an equivalent activity, and so on. In turn, scheduling requires new classes of market-like algorithms to ensure cooperation of multiple autonomous domains.

7.3 KOLMOGOROV COMPLEXITY AND THE COMPLEXITY OF MARKOV CHAINS

The question how to measure the descriptive complexity of an object was addressed by the great Russian mathematician Kolmogorov [3] and independently by Solomonoff [6] and Chaitin [1].

The description of an object is a string, henceforth, it seems quite appropriate to relate the complexity of the object with the length of the shortest string capable to describe the object. Intuitively, we realize that very long strings may have very short descriptions, while other descriptions are incompressible. For example, a binary string consisting of 10^{100} repetitions of the pattern 011 has a very short description while there is no shorter description of a string of the same length consisting of randomly generated 0's and 1's. It is interesting to note that the expected length of the shortest description of a random variable is approximately equal to its entropy.

This approach is intuitive and has been known for centuries. "Nunquam ponenda est pluritas sine necesitate", the famous principle formulated by William of Ockham (1290?-1349?), states that an explanation should not be extended beyond what is necessary [7]. Bertrand Russel translates this as "It is vain to do with more what can be done with fewer."

The Kologorov complexity $K_V(s)$ of the string s with respect to the universal computer V is defined as the minimal length over all programs $Prog_V$ that print s and halt:

$$K_V(s) = min[Length(s)] \quad over \ all \quad Prog: \ V(Prog_V) = s.$$

We consider only binary strings, but it is easy to see that the results apply to any alphabet. We now review several properties of Kolomgorov complexity.

Property 1 - Universality. If V is a universal computer, then for any other computer W:

$$K_V(s) \leq K_W(s) + constant_W.$$

Proof:

The basic idea of the proof is that we can construct a simulator of machine W running on our universal machine V, $Simulator_V(W)$. If we have a program for W capable to print s, $Prog_W(s)$, then we can supply to V a program called $Prog_V(s)$ consisting of $Simulator_V(W)$ and the program printing s on W, $W(Prog_W) = s$. Then:

$$Length[Prog_V(s)] = Length[Simulator_V(W)] + Length[Prog_W(s)].$$

Thus:

$$K_V(s) = min\{Length[Prog_V(s)]\} \leq K_W(s) + constant_W.$$

This result allows us to drop the reference to a particular machine in the expression of Kolmogorov complexity.

Let us call $K(s \mid Length(s))$ the conditional Kolmogorov complexity of string s given its length, $Length(s)$.

Property 2 - Upper and Lower Bounds of Kolmogorov complexity.

$$K(s \mid Length(s)) \leq Length(s) + constant.$$

$$K(s) \leq K(s \mid Length(s)) + log^*(Length(s)) + constant.$$

where

$$log^*(n) = log(n) + log(log(n)) + log(log(log(n))) + \dots$$

here we continue the sum until the last positive term.

The number of binary strings with complexity $K(s) < k$ is less than 2^k.

Proof:

If we know the length of the string s, then $K(s|Length(s))$, has the length itself as an upper bound. Indeed, we can embed the string itself in the program and the end of the program is well defined because we know the length of the string.

If we do not know the length of the string n we need to include it in the description. We first specify $\lceil log(n) \rceil$, the number of bits in the description of n and then we specify n. To specify the integer representing the number of bits in $\lceil log(n) \rceil$ we can use the same trick, determine the number of bits in the binary representation of $\lceil log(n) \rceil$ and then specify $\lceil log(n) \rceil$. We can continue this process until the last positive integer. Thus, we can describe n in $log^*(n)$ bits.

As far as the lower bound is concerned, we observe that there are $2^k - 1$ strings of length k or shorter. Indeed:

$$1 + 2^1 + 2^2 + \cdots + 2^{k-1} = 2^k - 1.$$

Property 3 - The Joint Kolmogorov Complexity.

Let s_1 and s_2 be two strings and denote by $K(s_1, s_2)$ the length of the shortest program that compute both s_1 and s_2 and is able to tell them apart. Then:

$$K(s_1, s_2) \leq K(s_1) + K(s_2) + 2 \times log[min(K(s_1), K(s_2))].$$

Proof:

Call $Prog(s_1)$ and $Prog(s_2)$ the shortest programs to produce s_1 and s_2 respectively. We need $\mathcal{O}(1)$ extra bits to schedule a new program, $Prog(s_1, s_2)$, capable of producing s_1 followed by s_2 using $Prog(s_1)$ and $Prog(s_2)$. The problem is now to identify s_1 and s_2 in the output. This can be done by using as input the shortest string of $[Length(s_1)]s_1 s_2$ or $[Length(s_2)]s_2 s_1$.

Here the notation $[n]$ means that we specify an integer n and a marker indicating the end of the binary representation. A solution requiring $2 \times log(n) + 2$ bits to specify an integer and its length is discussed next; we double each of the $log(n)$ bits in the binary representation of integer n, and add the string 10 as a marker to signal the end of the string. For example 10 has the binary representation 1010; it will be represented as 1100110010.

The joint Kolmogorov complexity of n objects $s_1, s_2, \ldots s_n$ has an upper bound:

$$K(s_1, s_2, \ldots s_n) \leq \sum_{i=1}^{n} K(s_i) + \mathcal{O}(log^* \sum_{i=1}^{n} K(s_i)).$$

The proof is by induction. This is certainly true for $n = 2$.

$$min(K(s_1), K(s_2)) \leq K(s_1) + K(s_2).$$

If we assume that it is true for n we need to prove that it holds for $n + 1$.

$$K(s_1, s_2, \ldots s_n, s_{n+1}) \leq K(s_1, s_2, \ldots s_n) + K(s_{n+1}) +$$

$$+ 2 \times log[min\{K(s_1, s_2, \ldots s_n), K(s_{n+1})\}] \leq \sum_{i=1}^{n+1} K(s_i) +$$

126

$$+ \mathcal{O}(log^* \sum_{i=1}^{n} K(s_i)) + 2 \times log[min\{[\sum_{i=1}^{n+1} K(s_i) + \mathcal{O}(log^* \sum_{i=1}^{n} K(s_i))], K(s_{n+1})\}].$$

Thus:

$$K(s_1, s_2, \ldots s_n, s_{n+1}) \leq \sum_{i=1}^{n+1} K(s_i) + \mathcal{O}(log^* \sum_{i=1}^{n+1} K(s_i)).$$

For specific values of n we can get tighter upper bounds: For example for $n = 3$:

$$K(s_1, s_2, s_3) \leq KS_3 + 4 \times log(KS_3 + log(KS_3)).$$

with:

$$KS_3 = K(s_1) + K(s_2) + K(s_3).$$

For $n = 4$

$$K(s_1, s_2, s_3, s_4) \leq KS_4 + 4 \times log(KS_4) + 2 \times log(KS_4 + 4 \times log(KS_4)).$$

with:

$$KS_4 = K(s_1) + K(s_2) + K(s_3) + K(s_4).$$

Recall that a stochastic process is an indexed sequence of random variables; a Markov process is a memoryless process. Let us turn our attention to the description of a stationary Markov process X with 2^n states labeled $X^i = (x_1^i, x_2^i, \ldots x_n^i)$, with $x_j^i = (0, 1)$ $1 \leq i \leq 2^n$, $1 \leq j \leq n$. The Markov chain has distribution μ and transition matrix P. We are now interested in the question on how the entropy of the sequence of n random variables $X^1, X^2, \ldots X^n$ grows with n.

Definition - The Entropy Rate of a Stochastic Process X:

$$H(\mathcal{X}) = \lim_{n \to \infty} \frac{1}{n} H(X^1, X^2, \ldots X^n).$$

where $H(X^1, X^2, \ldots X^n)$ is the joint entropy of the random variables. The entropy rate gives the entropy per symbol of the n random variables and it is equal with the conditional entropy of the last random variable given the past [2].

Theorem - The Entropy Rate of a Stationary Markov Chain.

Consider a stationary Markov chain with distribution μ and transition matrix P. Then according to [2] page 66 its entropy rate is:

$$H(\mathcal{X}) = - \sum_{i,j=1}^{n} \mu_i \times P_{i,j} \times log P_{i,j}.$$

127

We are now in the position to asses the complexity of the description of the Markov chain X. We encode the states of the chain using the Shannon code. Shannon code needs $log\frac{1}{p(X^i)}$ bits to describe state X^i where $p(X^i)$ is the probability of the system being in state X^i. Recall that the entropy of a discrete random variable X can be interpreted as the expected value of $log\frac{1}{p(X)}$, where X has the probability mass function $p(X)$ [2].

Thus we need $n \times H(X)$ bits to encode the states and we also need $\mathcal{O}(log(n))$ bits to describe the transitions:

$$K(X|n) \approx -n \times \sum_{i,j=1}^{n} \mu_i \times P_{i,j} \times logP_{i,j} + \mathcal{O}(log(n)).$$

Property - The Kolmogorov Complexity of a Stationary Markov Chain.

$$K(X) \le -n \times \sum_{i,j=1}^{n} \mu_i \times P_{i,j} \times logP_{i,j} + log^*(n) + \mathcal{O}(log(n)).$$

The proof of this property follows immediately from the basic properties of the Kolmogorov complexity.

7.4 A GRID COORDINATION AND SCHEDULING MODEL

In this section we model complex, long lasting computations that require a large and diverse set of resources. Therefore, static allocation of resources is impractical because each step may or may not be executed depending upon the input data; moreover, the precise resource needs for each step are known only approximately and static resource allocation based upon an off-line scheduling approach would be extremely wasteful.

Consider a computation $\mathcal{C}$ and a version of it, $\mathcal{C}^{grid}$, capable of runing on a computational grid. $\mathcal{C}^{grid}$ includes resource discovery, $\mathcal{D}$ and a transformation of computation $\mathcal{C}$ performing dynamic resource allocation, $\mathcal{C}^d$. From the expression of the joint Kolmogorov complexity it follows that:

$$K(\mathcal{C}^{grid}) = K(\mathcal{D}, \mathcal{C}^d) \le K(\mathcal{D}) + K(\mathcal{C}^d) + 2 \times log[min\{K(\mathcal{D}), K(\mathcal{C}^d)\}]. \quad (1)$$

To study $\mathcal{C}^d$ we assume that it requires m resources $r_1, r_2, \ldots r_j, \ldots r_m$. Call $n = 2^m$ and denote by $\rho = (\rho_1, \rho_2, \ldots \rho_j, \ldots \rho_n)$ the resource allocation vector; $\rho_j = 1$ if r_j is allocated and 0 otherwise. We say that the system is in state σ_i at time t if the resource allocation vector is ρ^i.

The system may only reach a subset of all possible states. For example, a computation may need four processors, $proc_1, proc_2, proc_3$ and $proc_4$; the resource allocation vector 1001 corresponds to a state when only processors $proc_1$ and $proc_4$ are allocated to the computation, and the system may never reach state 1011.

We assume that the states $\{\sigma_i\}$ form a stationary Markov chain with distribution μ and transition matrix P. Then the entropy rate of this Markov chain is:

$$H(\sigma) = -\sum_{i,j=1}^{n} \mu_i \times P_{i,j} \times logP_{i,j}. \quad (2)$$

Call

$$\beta = -n \times \sum_{i,j=1}^{n} \mu_i \times P_{i,j} \times logP_{i,j} + \mathcal{O}(log^*(n)). \quad (3)$$

The Kolmogorov complexity of the Markov chain σ with n states is:

$$K(\sigma) \leq \beta. \quad (3')$$

To describe $\mathcal{C}^d$ we need to describe the computations carried out in individual states, $\mathcal{S}$, as well as the Markov chain σ. Thus:

$$K(\mathcal{C}^d) = K(\mathcal{S}, \sigma) \leq K(\mathcal{S}) + K(\sigma) + 2 \times log[min\{K(\mathcal{S}), K(\sigma)\}]. \quad (4)$$

Now:

$$K(\mathcal{S}) = K(\sigma_1, \sigma_2, \ldots \sigma_n). \quad (5)$$

As shown earlier:

$$K(\sigma_1, \sigma_2, \ldots \sigma_n) \leq \sum_{i=1}^{n} K(\sigma_i) + \mathcal{O}(log^* \sum_{i=1}^{n} K(\sigma_i)). \quad (6)$$

In each state, σ_i, the system is expected to first execute a resource allocation procedure, $\mathcal{R}_i$ and then carry out the algorithmic component of $\mathcal{C}$ prescribed to that step, $\mathcal{A}_i$. Thus:

$$K(\sigma_i) = K(\mathcal{A}_i, \mathcal{R}_i) \leq K(\mathcal{A}_i) + K(\mathcal{R}_i) + 2 \times log\,(min\{K(\mathcal{A}_i), K(\mathcal{R}_i)\}). \quad (8)$$

$$\sum_{i=1}^{n} K(\sigma_i) \leq \sum_{i=1}^{n} K(\mathcal{A}_i) + \sum_{i=1}^{n} K(\mathcal{R}_i) + 2 \times \sum_{i=1}^{n} log\,(min\{K(\mathcal{A}_i), K(\mathcal{R}_i)\}). \quad (9)$$

or

$$\sum_{i=1}^{n} K(\sigma_i) \leq \sum_{i=1}^{n} K(\mathcal{A}_i) + \sum_{i=1}^{n} K(\mathcal{R}_i) + 2 \times log \prod_{i=1}^{n} min\{K(\mathcal{A}_i), K(\mathcal{R}_i)\}. \quad (10)$$

129

We consider the case when transformations of the algorithm to make it suitable for running on the grid are relatively simple; thus, the Kolmogorov complexity of all algorithmic steps can be approximated by the Kolmogorov complexity of the original computation

$$\sum_{i=1}^{n} K(\mathcal{A}_i) \approx K(\mathcal{C}). \quad (11)$$

We also assume that the pure algorithmic description at each step, $\mathcal{A}_i$, is more complex than the resource allocation procedure at that step, $\mathcal{R}_i$. Then:

$$\sum_{i=1}^{n} K(\sigma_i) \leq K(\mathcal{C}) + \sum_{i=1}^{n} K(\mathcal{R}_i) + 2 \times \log \prod_{i=1}^{n} K(\mathcal{R}_i). \quad (12)$$

The resource allocation procedure $\mathcal{R}_i$ is expected to determine if all resources needed in that state are already allocated and if not to use information gathered in the discovery phase to allocate it. Thus $\mathcal{R}_i$ may find itself in the states $\phi^i = \{\phi_1^i, \phi_2^i, \ldots \phi_k^i, \ldots \phi_{n_i}^i\}$ with $n_i = 2^i$.

We also assume that the states $\{\phi^i\}$ form a stationary Markov chain with distribution ν^i and transition matrix Q^i.

Call:

$$\alpha_i = -n_i \times \sum_{j,k=1}^{n_i} [\nu_j^i \times Q_{j,k}^i \times logQ_{j,k}^i] + \mathcal{O}(log(n_i)) + log^*(n_i). \quad (13)$$

$K(\mathcal{R}_i)$ is bounded:

$$K(\mathcal{R}_i) \leq \alpha_i. \quad (13')$$

Denote:

$$\alpha = \sum_{i=1}^{n} \alpha_i = -\sum_{i=1}^{n} 2^i \times \sum_{j=1}^{2^i} \sum_{k=1}^{2^i} \nu_j^i \times Q_{j.k}^i \times logQ_{k,j}^i + \mathcal{O}(n). \quad (14)$$

Then:

$$\sum_{i=1}^{n} K(\mathcal{R}_i) \leq \alpha. \quad (14')$$

It follows that:

$$\sum_{i=1}^{n} K(\sigma_i) \leq K(\mathcal{C}) + \alpha + 2 \times \log \prod_{i=1}^{n} \alpha_i. \quad (15)$$

130

$$K(\mathcal{S}) \leq K(\mathcal{C}) + \alpha + 2 \times log \prod_{i=1}^{n} \alpha_i + \mathcal{O}[log^*\{K(\mathcal{C}) + \alpha + 2 \times log \prod_{i=1}^{n} \alpha_i\}]. \quad (16)$$

Let us turn our attention to $K(\mathcal{C}^d)$ expressed by equation (4). We distinguish two cases.

Case 1 $K(\mathcal{S}) \leq K(\sigma)$. Then

$$K(\mathcal{C}^d) \leq K(\mathcal{S}) + 2 \times logK(\mathcal{S}) + \beta. \quad (17)$$

Case 2 $K(\mathcal{S}) > K(\sigma)$. Then

$$K(\mathcal{C}^d) \leq K(\mathcal{S}) + \beta + 2 \times log\beta. \quad (18)$$

Note that both α and β are monotonously increasing functions of n thus the complexity of $K(\mathcal{C}^d)$ increases with $n = 2^m$.

The resource discovery procedure can be modeled also as a stationary Markov chain with a number of states given by the number of distinct administrative domains. Arguments similar with the ones presented earlier allow us to conclude that the complexity of the resource discovery procedure $K(\mathcal{D})$ increases as the number of autonomous domains increases.

7.5 CONCLUSIONS

In addition to heuristics, algorithms, policies, mechanisms, and implementations we also need a framework to evaluate the limitations of current solution. The grid research community is still in the process of assessing the implications of the system complexity and the reasons why many scheduling proposals do not scale very well. For example, schemes involving the periodic monitoring of potential target systems cannot possibly work well at the scale of the entire grid.

In this paper we propose a model to study the complexity of scheduling applications consisting of dependent tasks on a computational grid. The model is consistent with our intuition that the finer the granularity of the resources allocated to each task, the more complex the tasking structure of the application, and the larger the number of autonomous domains involved, the more complex the scheduling problem becomes.

A practical conclusion is that service grids supporting a much coarser granularity of resource allocation, applications with a less complex tasking structure, and a lesser degree of coordination among autonomous domains are more feasible at this stage of grid computing than computational grids.

It is very likely that the efforts to build computational grids will be based on ad-hoc solutions working well for some very special cases.

7.6 ACKNOWLEDGEMENTS

The research reported in this paper was partially supported by National Science Foundation grants MCB9527131, DBI0296107, ACI0296035, and EIA0296179.

REFERENCES

1. G. J. Chaitin. *On the Length of Programs for Computing Binary Sequences*. J. Assoc. Comp. Mach. 13:547-569, 1966.

2. T. M. Cover and J. A. Thomas. *Elements of Information Theory*. Wiley, New York, NY, 1991.

3. A. N. Kolmogorov. *Three Approaches to the Quantitative Definition of Information*. Problemy Peredachy Informatzii, 1:4-7, 1965.

4. M. Li. and P. Vitany. *An Introduction to Kolmogorov Complexity and Its Applications*. Springer Verlag, Heidelberg, Second Edition, 1997.

5. D. C. Marinescu. *Internet-Based Workflow Management: Towards a Semantic Web*. Wiley, New York, NY, 2002.

6. R. J. Solomonoff. *A Formal Theory of Inductive Inference*. Inform. and Control 7(1-22): 224–254, 1964.

7. S. C. Tornay. *Ockham: Studies and Selections*. Open Court Publishers, La Salle, IL, 1938.

Author(s) affiliation:

- **Dan C. Marinescu, Gabriela M. Marinescu, and Yongchang Ji**

  School of Electrical and Computer Engineering
  University of Central Florida
  Orlando, Florida, 32816, USA
  Email: [dcm, magda, yji]@cs.ucf.edu

8

Programming the Grid with Distributed Objects

Alexandre Denis
Christian Pérez
Thierry Priol
André Ribes

Abstract

Computational grids are seen as the future emergent computing infrastructures. However, their programming requires the use of several paradigms that are implemented through several communication middleware and runtimes (distributed objects, message-passing libraries, ...). However some of these middleware and runtimes are unable to take benefit of the presence of various networking technologies available in grid infrastructures. In this paper, we propose an approach that is capable of exploiting the underlying networking technologies at their maximum performance. This approach is based on the CORBA architecture from the OMG for which we propose the concept of parallel CORBA object able to exploit Wide Area Networks (WAN) technologies and a supportive environment able to exploit various System Area Networks (SAN) technologies. Such approach encourages grid programmers to use the most suited communication paradigms for their applications independently from the underlying networks.

8.1 INTRODUCTION

With the availability of high-performance networking technologies, it is nowadays feasible to couple several computing resources together to offer a new kind of computing infrastructure that is called a Computational Grid [7, 8]. A Computational Grid acts as a high-performance virtual supercomputer to users to perform various applications such as for scientific computing or for data management. Building Computational Grids raises the same design issues as for distributed systems: transparency (location of resources is transparent to the user), interoperability (to hide the heterogeneity of computing and networking resources) and reliability (the system has to survive the unavailability of computing and networking resources). It also shares the same design issues as for parallel systems: performance (best use of both computing and networking resources) and scalability (efficient management of a huge number of resources).

Software infrastructures, such as Globus[7] or Legion[9], aim at providing runtime systems to allow the execution of applications on Computational Grids. However, Globus was mainly designed to allow the execution of parallel applications. Such approach makes senses since there are already a huge number of existing parallel applications that should benefit from Computational Grids. However, the availability of Computational Grids will give rise to new kind of applications for which parallel programming, based on the use of message-passing libraries, is not suitable. Coupled simulations are an example of such new kinds of application. It aims at coupling several parallel codes to simulate complex systems that require a multi-physics approach. Therefore, one important question arises when using a grid system: what is the most appropriate approach to program a Computational Grid, or said differently, what programming models have to be provided to Grid application designers ? On that matter, there is no consensus mainly due to the wide nature of applications that could benefit from Computational Grids. Since such systems are a combination of parallel and distributed systems, it is very tempting to extend programming models that were associated to parallel systems (message passing libraries, shared memory) so that they can be used for distributed programming. Similarly, programming models for distributed systems (remote procedure call, distributed objects) can be adapted to program parallel systems. Neither of these two approaches can be seen as viable solutions for the future of Grid Computing. It is thus important to try to combine the two different worlds into a single coherent one. Our objective is to design a programming environment for the Grid based on the use of distributed objects (for code coupling) and message-passing (for parallel processing within a scientific code). This objective is challenging if performance is wanted. First of all, encapsulation of parallel codes into objects should be done in such a way that communication between objects is scalable. Secondly, distributed objects have to communicate through various system area networking technologies whereas most of the middleware implementations associated with distributed objects are only based on TCP/IP. This paper will show solutions to solve these problems in the context of the CORBA architecture.

Among a large set of distributed programming technologies, CORBA is probably the most promising one due to its object oriented approach and its independence from

```
#include "Matrix.idl"

interface IExample {
    void send_data(Matrix m);
}
```

Fig. 8.1 IDL interface of the parallel object

```
void f(long* A, int size) {
    IExample obj("Servant");
    Matrix<long> data(1);          // create a Matrix of 1 dimension
    data->setBounds(0,1,size);    // bounds [1,size[ for dimension 0
    data->setData(A);             // initialize data pointer (no data copy)
    obj->send_data(data);         // remote invocation
}
```

Fig. 8.2 Motivating Example: a sequential client calls a parallel method.

operating systems and languages. CORBA is a specification from the OMG (Object Management Group) to support distributed object-oriented applications. CORBA acts as a *middleware* that provides a set of services allowing the distribution of objects among a set of computing resources connected to a common network. Transparent remote method invocations are handled by an Object Request Broker (ORB) which provides a communication infrastructure independent of the underlying network. An object interface is specified using the Interface Definition Language (IDL) that gives a list of allowed operations on a particular object. As a distributed programming technology, CORBA can be used as a "glue" to couple several high-performance simulation codes that are executed on different computing resources of a computational Grid. However, CORBA lacks of supporting efficiently the encapsulation of parallel codes and most of the ORB implementations only provide support for TCP/IP avoiding the use of system area networking technologies.

The remainder of this paper is divided as follows. Section 8.2 presents a solution to encapsulate parallel codes into CORBA objects in such a way that communication between objects, encapsulating parallel codes, is scalable. Section 8.3 sketch the architecture of a communication framework that is able to let communication middleware and runtimes to exploit system area networks. Section 8.4 presents some concluding remarks.

8.2 USING WIDE AREA NETWORKS WITH CORBA OBJECTS

CORBA lacks of supporting efficiently the encapsulation of high-performance simulation codes that were designed to be run on parallel systems. Encapsulation of parallel codes (we assume that they follow an SPMD execution model) with standard CORBA implementations requires selecting one SPMD process to be encapsulated within a CORBA object. This particular process acts as a master to drive the execution of the other SPMD processes (slaves). Synchronization and data transfer between the master and slaves processes go through a message-based communication layer

135

such as MPI. Such approach does not allow all the SPMD processes to be connected to the ORB. Communication from a client to the server has to go through a single SPMD process (the one encapsulated into the object implementation). Such approach does not offer a scalable way to communicate between the client and the server. We can imagine a situation where the client itself is a parallel code encapsulated into a CORBA object running on a PC cluster trying to send a very large matrix to another CORBA object that encapsulates a parallel code running on another PC cluster. The communication between the client and the server will be bound by the performance of the networking interface of the machine that run the SPMD code encapsulated into the CORBA object even if the network that connects the two PC clusters is a WAN with a higher bandwidth.

To overcome this problem, we introduced the concept of Parallel CORBA object. It is defined as a collection of identical CORBA objects. Each CORBA object encapsulates an instance of the SPMD code (a SPMD process) allowing an SPMD parallel code to communicate with the external world through several communication channels. Therefore, internal communication, within the SPMD code, is carried out through message-based libraries (such as MPI) and external communication (for code coupling) is done thanks to the CORBA ORB. Our objective is to hide as much as possible the collection of CORBA objects to the users. In [12], we proposed an implementation that requires a modification to the IDL language. This work targets parallel CORBA objects on top of compliant CORBA ORBs without involving whatsoever modification of the CORBA specifications. We call such objects portable parallel CORBA objects. Throughout this section, we discuss with respect to a motivating example.

8.2.1 Motivating Example

Figure 8.1 presents the user level IDL interface of the motivating example presented in Figure 8.2. A sequential client wants to send an array A to a method void send_data(Matrix m) of the interface IExample. The client knows that this service is implemented by an object of named Servant. But, the client does not know – and does not want to know – that the implementation is in fact parallel. To connect to the object, the client instantiates a local object obj of type IExample with the name of the remote object as argument. Then, once the Matrix view of its local array A is built, the method is invoked.

8.2.2 Achieving Portable Parallel CORBA Objects

To implement this kind of example on top of a compliant CORBA ORB, we need to introduce a layer between the user code and the ORB, as depicted in Figure 8.3. This layer embeds the complexity of connection and data distribution management. Its main role is to map an user-level interface – IExample in the example – to an IDL interface, that is called ManagerIExample. This latest interface contains the methods defined by the user as well as private methods. The private methods provide

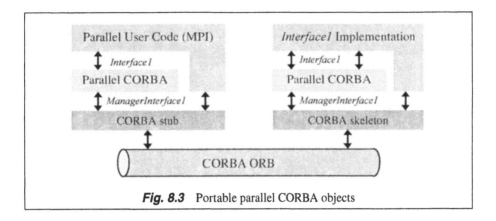

Fig. 8.3 Portable parallel CORBA objects

services like the localization of all remote objects being part of the implementation of IExample and the retrieval of the data distribution of arguments of user-level methods. The client and server side methods of the parallel CORBA object layer are analog to the stub and the skeleton of ORB requests. But, while stubs and skeletons of ORB requests deal with peer-to-peer issues (like data marshaling), the stub and skeletons of the parallel CORBA object layer concentrate on data distribution issues. Finally, the stubs and the skeletons of the parallel CORBA layer should be generated from an IDL level description of the user services. However, they are currently hand-written. The rest of this section reviews different aspects of the internals.

Connection Management. A parallel object is defined by a name (string). This name in fact represents a context in the Naming Service that contains two kind of entries: the IOR of the service manager and all the IOR of the objects that belongs to the parallel objects, as illustrated in Figure 8.4. The constructor of IExample retrieves information like the number of objects thanks to the Manager object. Then, it can collect their respective IOR from the Naming Service.

Method Invocation. When the client invokes the send_data method, it in fact calls the corresponding method of the ManagerIExample interface, locally implemented into the parallel CORBA layer. This method builds CORBA requests according to the data distributions expected by the parallel objects. Such information is available thanks to methods belonging to the ManagerIExample Interface. Then, it sends the CORBA requests to the ManagerIExample objects. The role of the server side method is to gather data coming from different clients (when the client is parallel) before calling the server side implementation of the send_data method. Similarly, it scatters the out arguments.

When a client invokes a method of a parallel object, it potentially has to send several CORBA requests. An efficient and reliable solution would be the use of the Asynchronous Message Interface that appears in CORBA 2.4. As we were not aware of open source ORB that supports this feature, we implemented a temporary solution based on *oneway* requests. This solution has severe limits. First, it is not a

137

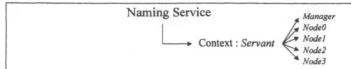

Fig. 8.4 A parallel CORBA object of name *Servant* registers all its CORBA objects in the naming service under the context of same name. This context also contains the IOR of the manager and the IOR of all objects.

reliable solution as such kind of requests are not reliable according to the CORBA specifications. But, as we used TCP to transport CORBA requests, all *oneway* requests are delivered. Second, we have to build a system to detect the termination of the request.

Data Distribution Management. The core of parallel objects is the data distribution management. From our experience, mainly derived from previous works [12] and High Performance FORTRAN[6], we believe its important to have a high level of transparency: our choice is to separate the data distribution from the interface. By decoupling the data distribution from the interface, we obtain four major benefits. A first benefit is there is no need to modify the CORBA IDL. The second benefit is that argument data distribution is transparent to the user, as distribution does not appear in the interface. A third benefit is that a parallel object can dynamically change the distribution pattern it is awaiting. This may happen for example if some objects are removed (due to node failure for example) or some objects are added. This feature implies some interesting issues. For example, how is the client informed? A solution would be to use a listener design pattern. A second issue is: what does a parallel object do with incoming requests that have an argument with an old distribution? If all the data has correctly been received, a redistribution may be performed. However, whenever some data is missing (node failure) or the parallel object does not implement the redistribution feature, a CORBA exception is returned to the client. The fourth benefit is the ease of the introduction of new data distribution patterns as only clients and parallel objects that use non standard data distributions have to know about these.

Intermediate Matrix Type. Applications are expected to be written with their own data distribution scheme. So, we face the problem of embedding user data into a standard IDL representation so as to provide interoperability. We achieve data distribution interoperability thanks to a `Matrix` interface, sketched in Figure 8.5. It provides a logical API to manipulate an internal IDL representation of data distributions. This API should be straightforward for client (like in the example of Figure 8.2) and should provide functionalities for implementers. Internally, the `Matrix` interface manages an IDL structure that contains distribution information as well user data. That is this structure which is sent through the ORB. Currently, we only implement the `Matrix` interface as a C++ class whose API provides methods that manages a C++ repre-

138

```
interface Matrix {

  struct dim_t { long  size,  low,  high; };

  struct matrix {
    dis_t              dis;    // current distribution
    long               ndim;   // number of dimension
    sequence<dim_t>    rdim;   // global view of the array
    sequence<dim_t>    ddim;   // local  view of the array
    data_t             data;   // data
  };
};
```

Fig. 8.5 IDL distributed array representation

```
Matrix<float> data(2);              // matrix with 2 dimension

data.setBounds(0,0,size1);          // Set bounds for dimension 0
data.setBounds(1,0,size2);          // Set bounds for dimension 1
Distribution d0(Matrix::BLOCK, procid, nbproc);
Distribution d1(Matrix::SEQ);
data.setDistribution(0, d0);        // Set distribution for dimension 0
data.setDistribution(1, d1);        // Set distribution for dimension 1
data.allocateData();                // Allocate memory

for( int i0 = data.low(0); i0 < data.high(0); i0++ )
  for( int i1 = data.low(1); i1 < data.high(1); i1++ )

    data(i0, i1) = ...
```

Fig. 8.6 C++ server side example: initialization of a 2D distributed array of floats which
has a block-distributed dimension. i0 and i1 are global indexes.

sentation of the IDL `Matrix` structure. While Figure 8.2 has provided a client side
example, Figure 8.6 presents a server side example that illustrates the initialization
of a 2D distributed array.

8.2.3 Experiments

Very recently, we had access to the VTHD network. It's an experimental network of
2.5 Gb/s that in particular interconnects two INRIA laboratories, which are about one
thousand kilometers apart. The test platform is made of two PC clusters connected
through the VTHD network. The network interface of each node of these two clusters
is a standard Fast Ethernet (100 Mb) and the communication protocol is TCP/IP.
We made our experiments with OmniORB 3 [1]. In a peer-to-peer situation using
OmniORB we measure a throughput of 11 MB/s; the Ethernet 100 Mb/s card being the
limiting factor. For experiments with an 8-node parallel client, running on one cluster,
and an 8-node parallel object, running on the other cluster, we measure an aggregated
bandwidth of 85.7 MB/s, which represents a point-to-point bandwidth of 10.7 MB/s.
Portable CORBA parallel objects prove to efficiently aggregate bandwidth and thus
it is capable of exploiting a WAN in a scalable way.

139

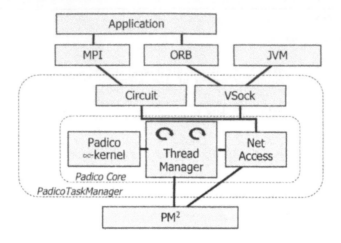

Fig. 8.7 PadicoTM architecture.

8.3 USING SYSTEM AREA NETWORKS WITH CORBA OBJECTS

The design of coupled applications based on CORBA raises the issue of supporting this middleware on various system area networking technologies. One can imagine that such applications have to be executed on a single PC cluster (depending of the availability of computing resources within a Grid). Such cluster might be equiped with a very efficient SAN such as Myrinet or SCI. In such a situation, CORBA cannot take benefit of such a network since most of the actual implementations rely only on TCP/IP. We thus envisaged to design a supportive environment allowing CORBA to exploit such networking technologies. The problem is not as simple as it seems: an implementation of a specific protocol within a CORBA ORB. Indeed, within a parallel CORBA object, communication is carried out by a message-passing library (for synchronization and data redistribution). It is thus required that such supportive environment must support several communication middleware and runtimes to multiplex the network interface. As for instance, supporting CORBA and MPI, *both running simultaneously*, is not straightforward. Several access conflicts for networking resources may arise. For example, only one application at a time can use Myrinet through BIP [11]. If both CORBA and MPI try to use it without being aware of each other, there are access conflicts and reentrance issues. If each middleware (eg. CORBA, MPI, etc.) has its own thread dedicated to communications, with its own policy, communication performance is likely to be sub-optimal. If ever we are lucky enough and there is no resource conflict, there is probably a more efficient way than putting side by side pieces of software that do not see each other and that act in an "egoistic" fashion. In a more general manner, resource access should be cooperative rather than competitive. PadicoTM is our research platform to investigate the problems of integrating several communication middleware and runtime (not only CORBA and MPI). The following sections give a description of this platform.

140

8.3.1 PadicoTM Overview

The design of PadicoTM, derived from the software component technology, is very modular. Every module is represented as a component: a description file is attached to the binary files. PadicoTM is composed of *core* modules and *services* plugged in the core. PadicoTM core implements module management, network multiplexing and thread management. PadicoTM core comprises three modules: *Puk*, *TaskManager* and *NetAccess*. Services are plugged in PadicoTM core. The available services are:

- advanced network API (*VSock* described in Section 8.3.5 and *Circuit* described in Section 8.3.6) on top of native PadicoTM network API;

- middlewares and runtimes, namely a CORBA module (Section 8.3.8.2), a MPI module (Section 8.3.8.1) and a Java Virtual Machine;

- gatekeepers (Section 8.3.7) which enable the user to remotely steer the processes on each node.

Currently, we have a functional prototype with all these modules available.

8.3.2 Dynamicity

There is a network model discrepancy between the "distributed world" (eg. CORBA) and the "parallel world" (eg. MPI). Communication layers dedicated to parallelism typically use a static topology[1]: nodes cannot be inserted or removed into the communicator while a session is active. On the other hand, CORBA has a distributed approach: servers may be dynamically started, clients may dynamically contact servers. The network topology is dynamic. High-performance networks API are mostly biased toward the parallel model; thus, it is challenging to map the distributed communication model of CORBA onto SAN. Since most communication libraries for SAN (eg. BIP, vendor's MPI on most machines, Madeleine [2]) require the processes on all nodes to be started at the same time, we chose that PadicoTM bootstraps a unique binary on each node. It satisifies the SPMD requirement of the communication library. Since we do not want all nodes to run the same application, we chose to put applications in *loadable modules*. The user's applications are stored into dynamically loadable modules. Thanks to this mechanism, different binaries can dynamically be loaded into the different nodes of a cluster or parallel computer. For example, we can load a CORBA server on one node and CORBA clients on the other nodes. In PadicoTM, we call this bootstrap binary *Padico μ-Kernel*, or in shorter *Puk*. Once the *Puk* is bootstrapped on each node, it loads the other modules and starts them.

We want the *module* concept to be open. We do not restrict ourselves to binary dynamically loadable libraries. Actually, modules are described in a file written in XML. This description file contains: the name of a *driver* able to load this module, references to other modules for dependency checking, *units* and attributes. A driver is

---

[1]PVM and MPI2 address this problem but do not allow network management on a link-per-link basis.

```
<mod name="ORB" driver="binary">
  <requires>VSock</requires>
  <attr label="NameService">
    corbaname::paraski:10000
  </attr>
  <unit>libORB.so</unit>
</mod>
```

Fig. 8.8 The XML description for the *ORB* service.

a set of functions which tell *Puk* how to load, start and unload a type of unit. Different drivers may be seen as module types. For example, there is the binary driver which defines units as binary shared objects (".so" libraries on Unix systems), the java driver which defines units as Java classes, or the pkg driver which defines units as being modules. Attributes are environment variable aimed at configuring modules. Figure 8.8 is the XML description for the *ORB* module: it should be loaded by the binary driver, requires the *VSock* module, contains the libORB.so unit and an attribute for referencing the CORBA name service.

8.3.3 Thread Management

It is now common that middleware use multi-threading. However, middlewares which are not designed to run together in the same process are likely to use incompatible thread policies, or simply different multi-threading packages. An application runs into trouble when mixing several kinds of threads. That is why PadicoTM must provide the plugged-in middleware with a portability layer for multi-threading. At first look, it may seem attractive to use Posix threads (known as pthread) as a foundation. However, it has been shown [4] that MPI and Posix threads do not stack up nicely. To deal with portability as well as performance issues, we choose the Marcel [5] multi-threading library. Marcel is a multi-threading library in user space. It implements an N:M thread scheduling on SMP architectures. Marcel has been designed to guarantee a good reactivity of the application to network I/O when used in conjunction with the Madeleine [2] communication layer. The *TaskManager* module of PadicoTM is based on Marcel. Every PadicoTM modules which use multi-threading are supposed to use Marcel and no other multi-threading library. This is not very constraining: Marcel API is very similar to Posix threads.

8.3.4 Cooperative Access to System Area Networks

Access to high speed networks is the more conflict-prone task when using multiple middleware at the same time. Some access methods require an exclusive access to the hardware (eg. Myrinet through BIP) thus only one library can use it at the same time – CORBA *or* MPI, not both; some networks have limited resource which can be exhausted if different libraries open separate connections (eg. SCI); some network

142

hardware can be used through several drivers which cause conflicts if used at the same time by different middleware – if middleware use Myrinet, they must agree on the driver to use: BIP *or* GM, not both at the same time. In the worst case, middleware cannot coexist in the same process nor on the same machine, due to network access conflict. In the best case, if the middleware do not know each other, each middleware would run its own polling thread so that the access to the network is competitive and prone to race conditions.

To deal with low level, portability, and performance issues, we chose to use Madeleine [2] as a foundation for the *NetAccess* module of PadicoTM. The Madeleine communication layer was designed to bridge the gap between low-level communication interfaces (such as BIP [11], SBP or UNET) and middleware. It provides an optimized interface for *RPC-like* operations that allows zero-copy data transmissions on high-speed networks such as Myrinet or SCI, and is best used with Marcel threads. A unique polling loop managed by the PadicoTM *NetAccess* module dispatches incoming messages to modules that want access to high-speed networks. Thus, every modules use the network through Madeleine: there are no access conflict. Moreover, there is no competition thanks to the unique polling loop.

In order to allow several middleware to use the network through Madeleine, there is a need for multiplexing in some layer. Madeleine itself can multiplex messages, but it is limited by the hardware. For example, Madeleine provides two channels on top of BIP, and only one channel on top of SCI. However, we want to be able to deploy an arbitrary number of networking middleware in a PadicoTM process, so we need an arbitrary number of logical communication channels. The *NetAccess* module needs to multiplex logical "PadicoTM channels" on top of Madeleine hardware channels. Practically, *NetAccess* uses one Madeleine channel with one polling loop listening on it. The modules which want to use Madeleine register callback functions which are called when a message arrive. To guarantee that the communications are deadlock-free, callbacks are not allowed to block nor to send directly a message on the network. However, if they need to send a reply or to wait on a condition, the *TaskManager* can do it in another thread.

Puk, *TaskManager* and *NetAccess* modules are PadicoTM core. Other modules are called services. They are plugged in the PadicoTM core. Figure 8.9 sums up the different available modules in PadicoTM.

8.3.5 Virtual Sockets

The TCP/IP network protocol is designed for use over LAN or WAN. It is not well suited for use over a SAN. Moreover, system calls add a significant latency to the data path. That is why we avoid as much as possible kernel-level communication libraries. However, the widespread socket interface from Berkeley is fairly well suited for networking. Most networking middleware use sockets; some of them heavily rely on the concept of sockets and would require very deep changes to use another communication paradigm. Thus, we chose to implement a socket-like interface on top of the "native" *NetAccess* interface described in the previous section, like Fast Socket [13] on top of Active Messages. Our approach relies on the concept of *virtual*

143

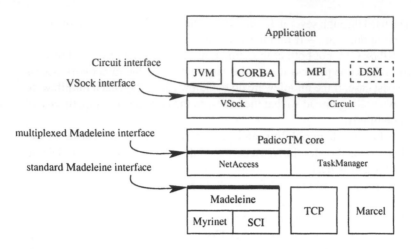

Fig. 8.9 PadicoTM modules

socket, that we call *VSock*. *VSock* implements a subset of the standard socket functions in user space on top of *NetAccess*, for achieving high-performance. It performs zero-copy datagram transfer with a socket-like connection handshake mechanism and it offers a multi-protocol communication layer with auto-selection. It automatically selects the adequate protocol to use according to the available hardware.

8.3.6 Groups and Circuits

The *NetAccess* module is a low-level communication layer of PadicoTM. It creates communication channels which comprise every nodes of a cluster. However, one may want for example to deploy two MPI codes coupled with CORBA on a cluster. In this case, each MPI code spans across only a group of nodes, though the low-level communication library spans across all nodes.

To handle such cases, PadicoTM provides the concept of logical groups of nodes. A group is a set of nodes of a cluster. We define a *circuit* as a *NetAccess* communication channel restricted to a group. Thus, higher level communication libraries, such as MPI, run on a circuit. The logical topology has not to match the hardware topology. This is different from creating MPI groups inside the high-level MPI communicator: there is no need to change an existing application which expect to use MPI_COM_WORLD, the middleware library (eg. MPI) is loaded only on nodes which actually run an MPI application. To manage modules on groups and circuits, we provide an additional driver for *Puk* called multi. The multi driver is aimed at running SPMD codes and SPMD middleware (such as MPI) on PadicoTM groups and circuits.

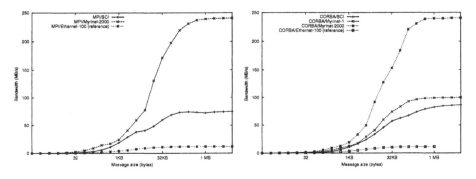

Fig. 8.10 MPI bandwidth   **Fig. 8.11** CORBA bandwidth

8.3.7 Remote Control

For dynamically monitoring and managing modules on each node, Padico comprises *PadicoControl*, a set of applications to remotely steer a PadicoTM process. Currently, there are two such applications: a GUI written in Java for portability, and a command-line tool for more advanced users. They may use two different contact methods: CORBA or XML-RPC. A PadicoTM service called *gatekeeper*, loaded in PadicoTM processes, listens to incoming requests and handles them (for example, load a module, return the list of running modules, etc.). It is mostly a remote interface for the *TaskManager* (see Section 8.3.3).

8.3.8 Experiments with Middleware and Runtimes with PadicoTM

PadicoTM comprises middleware that have been ported on top of PadicoTM services. The MPI implementation in PadicoTM is derived from MPICH/Madeleine [3] with very few changes (use *Circuit* instead of Madeleine and replace the polling thread with a callback). The CORBA implementation in PadicoTM is based on OmniORB3 [1] from AT&T. The porting of OmniORB on top of *VSock* and Marcel threads is straight-forward. The Java Virtual Machine module is based on Kaffe [10], on top of *VSock* and Marcel. Our benchmark machines are "old" dual-Pentium II 450 machines, with Ethernet-100, SCI and Myrinet-1, and "up to date" dual-Pentium III 1GHz with Myrinet-2000.

8.3.8.1 MPI The MPI module in PadicoTM gets the bandwidth shown on Figure 8.10. The peak bandwidth is excellent: 75 MB/s on SCI and 240 MB/s on Myrinet-2000. The latency is 23 μs on SCI and 11 μs on Myrinet-2000. This performance is very similar to MPICH/Madeleine [3] from which PadicoTM MPI implementation is derived; PadicoTM adds no noticeable overhead neither for bandwidth nor for latency.

8.3.8.2 CORBA The bandwidth of the high-performance CORBA implementation is shown on Figure 8.11. The benchmark consists in a remote invocation of a method which takes an *inout* parameter of variable size (sequence of long). The peak bandwidth is 89 MB/s on SCI, 101 MB/s on Myrinet 1, and 240 MB/s on Myrinet-2000. This performance is very good. We reach 99 % of the maximum achievable bandwidth with Madeleine.

On the "old" machines (Pentium II 450, SCI or Myrinet-1), the latency of CORBA for an empty remote invocation is around 55 μs. It is a good point when compared to the 160 μs latency of the ORB over TCP/Ethernet-100. On the "up to date" machines (Pentium III 1GHz, Myrinet-2000), the latency of CORBA is 20 μs where MPI gets 11 μs.

CORBA is as fast as MPI regarding the bandwidth, and slower than MPI for latency. This latency could be lowered if we used a specific protocol (called ESIOP) instead of the all-purpose GIOP protocol in the CORBA implementation. This performance is very good, though. As far as we known, this is the fastest CORBA implementation.

8.4 CONCLUSIONS

In this paper, we have described an extension and a supportive environment for the CORBA architecture that target coupled applications for computational Grids. The general belief, when promoting the use of CORBA in high-performance computing, is that such technology suffer from huge overhead and thus it is inappropriate. What we have shown in this paper is that such overhead is not intrinsic to the CORBA architecture but only to its implementation. Proper extensions to CORBA, without modifying the standard, can cope with the encapsulation of parallel codes in an efficient way. Moreover, communication through a CORBA ORB is always associated with high latency and low bandwidth. The PadicoTM platform shows that a CORBA ORB provides roughly the same bandwidth than MPI on a given networking technology and the latency is only twice the one with MPI. For these reasons, we think that the CORBA architecture is well suited to program computational grids, which is no wonder if we consider that a computational grid is simply a kind of distributed system. Of course, we do not solve all the problems with the current PadicoTM platform. Deployment and security issues remain but we think that such platform could be integrated with available grid software platform such as Globus.

REFERENCES

1. AT&T Laboratories Cambridge. OmniORB Home Page. http://www.omniorb.org.

2. O. Aumage, L. Bougé, J.-F. Méhaut, and R. Namyst. Madeleine II: A portable and efficient communication library for high-performance cluster computing. *Parallel Computing*, March 2001. To appear.

3. O. Aumage, G. Mercier, and R. Namyst. MPICH/Madeleine: a true multi-protocol MPI for high-performance networks. In *Proc. 15th International Parallel and Distributed Processing Symposium (IPDPS 2001)*, San Francisco, April 2001. IEEE. To appear.

4. L. Bougé, J.-F. Méhaut, and R. Namyst. Efficient communications in multi-threaded runtime systems. In *Parallel and Distributed Processing. Proc. 3rd Workshop on Runtime Systems for Parallel Programming (RTSPP '99)*, volume 1586 of *Lect. Notes in Comp. Science*, pages 468–482, San Juan, Puerto Rico, April 1999. In conj. with IPPS/SPDP 1999. IEEE TCPP and ACM SIGARCH, Springer-Verlag.

5. V. Danjean, R. Namyst, and R. Russell. Integrating kernel activations in a multi-threaded runtime system on Linux. In *Parallel and Distributed Processing. Proc. 4th Workshop on Runtime Systems for Parallel Programming (RTSPP '00)*, volume 1800 of *Lect. Notes in Comp. Science*, pages 1160–1167, Cancun, Mexico, May 2000. In conjunction with IPDPS 2000. IEEE TCPP and ACM, Springer-Verlag.

6. High Performance Fortran Forum. *High Performance Fortran Language Specification*. Rice University, Houston, Texas, October 1996. Version 2.0.

7. I. Foster and C. Kesselman. Globus: A metacomputing infrastructure toolkit. *The International Journal of Supercomputer Applications and High Performance Computing*, 11(2):115–128, Summer 1997.

8. I. Foster and C. Kesselman, editors. *The Grid: Blueprint for a New Computing Infracstructure*. Morgan Kaufmann Publishers, Inc, 1998.

9. A. S. Grimshaw, W. A. Wulf, and the Legion team. The Legion Vision of a Worldwide Virtual Computer. *Communications of the ACM*, 1(40):39–45, January 1997.

10. Kaffe: an OpenSource implementation of a Java Virtual Machine. http://www.kaffe.org.

11. L. Prylli and B. Tourancheau. Bip: a new protocol designed for high performance networking on myrinet. In *1st Workshop on Personal Computer based Networks Of Workstations (PC-NOW '98)*, Lect. Notes in Comp. Science, pages 472–485. Springer-Verlag, apr 1998. In conjunction with IPPS/SPDP 1998.

12. C. René and T. Priol. MPI code encapsulating using parallel CORBA object. In *Proceedings of the Eighth IEEE International Symposium on High Performance Distributed Computing*, pages 3–10, August 1999.

13. Steven H. Rodrigues, Thomas E. Anderson, and David E. Culler. High-performance local area communication with fast sockets. In *USENIX '97*, pages 257–274, January 1997.

Author(s) affiliation:

- **Alexandre Denis, and André Ribes**

  IRISA/IFSIC
  Campus Universitaire de Beaulieu
  35042 Rennes Cedex, France

- **Christian Pérez, and Thierry Priol**

  IRISA/INRIA
  Campus Universitaire de Beaulieu
  35042 Rennes Cedex, France
  Email: Thierry.Priol@inria.fr

9

Dynamic Data Driven Application Systems

"Creating a dynamic and symbiotic coupling of application/simulations with measurements/experiments"

Frederica Darema

Abstract

The novel capabilities to be created in the Dynamic Data Driven Applications Systems (DDDAS) context, are applications/simulations that can dynamically accept and respond to measurement-data dynamically injected into the application at execution time, and ability of the executing applications to dynamically control such measurements. This new paradigm has the potential to transform the way science and engineering are done, and induce a major beneficial impact in the way many functions in our society are conducted, such as manufacturing, commerce, transportation, hazard prediction/management, and medicine, to name a few. To achieve such goals, fundamental and multidisciplinary collaborative research is needed in applications, in mathematical algorithms, and in systems software, to enable DDDAS capabilities.

Traditional application simulations are conducted with static data inputs. The simulation executes to completion with that set of inputs, and further analysis of the system entails re-launching the simulation with a new set of inputs. In the new dynamic, data driven application systems envisioned here, field collected data will be used in an "online" fashion to steer the simulations. Reversely, the simulations could be used to dynamically control experiments or other field measurements. Thus the applications/simulations and the measurement processes become a symbiotic feedback system rather than the usual static, disjoint and serialized approaches existing today.

The DDDAS paradigm will enable a synergistic and symbiotic feedback control loop between applications/simulations and measurements. This novel technical direction that can open new domains in the capabilities of simulations with high potential payoff, and create applications and measurements with new and enhanced capabilities, and lead to breakthroughs in many application areas. Many of these applications are of interest to the research community supported by NSF.

This new paradigm creates a rich set of new challenges and new class of problems for the applications, algorithms, and systems researchers to address. Such challenges include: advances at the applications level for enabling this kind of dynamic feedback and control loop; advances in the applications algorithms for the algorithms to be amenable to perturbations by the dynamic data inputs and enhanced capabilities for handling uncertainties in input data; new technology in the computer software systems areas to support such environments. NSF organized a workshop to examine the technical challenges and research areas that need to be fostered to enable such capabilities. Representative examples of applications were addressed at the NSF workshop to illustrate the potential impact that this kind of research can have.

The NSF-wide Program ITR (Information Technology Research) with its broad scope has been used as a vehicle to start funding DDDAS-related research. Todate over a dozen such projects have been funded. While the community has expressed the view that a program focused on systematically funding research on DDDAS will expedite progress in the related technologies, ITR will continue as a venue for funding DDDAS- research.

http://www.cise.nsf.gov/eia/dddas/; includes Workshop Report

Author(s) affiliation:

- **Frederica Darema**

  National Science Foundation
  Arlington, Virginia 22230, USA
  Email: fdarema@nsf.gov

10

Towards Coordinated Work on the Grid: Data Sharing Through Virtual File Systems

Renato J. Figueiredo
Nirav Kapadia
José A. B. Fortes

Abstract

This paper describes a virtual file system that allows data to be shared among computational grid users distributed across multiple administrative domains. The virtual file system employs software proxies to broker transactions between standard Network File System (NFS) clients and servers; these proxies are dynamically configured and controlled by grid middleware. This technique enables data sharing at the granularity of files, providing a basis for grid-based coordination. The solution works with unmodified applications (even commercial ones) running on standard operating systems and hardware. Therefore, in addition to sharing, the virtual file system can leverage complementary coordination capabilities from legacy solutions based on NFS. Experimental results show that the wide-area performance of the current implementation of the virtual file system depends on application I/O requirements: it is within 1% of native local-area NFS performance for a typical compute-intensive PUNCH application (SimpleScalar), and 5.5 times worse than the local-area performance for an I/O-intensive application (Andrew).

10.1 INTRODUCTION

Network-centric computing promises to revolutionize the way in which computing services are delivered to the end-user. Analogous to the power grids that distribute electricity today, *computational grids* will distribute and deliver computing services to users anytime, anywhere. Corporations and universities will be able to out-source their computing needs, and individual users will be able to access and use software via Web-based computing portals.

A computational grid brings together *computing* nodes, *applications*, and *data* distributed across the network to deliver a network-computing session to an end-user. This paper elaborates on mechanisms to support data sharing across resources of a computational grid. The mechanisms are based on the abstraction of *logical user accounts* and the implementation of a *virtual file system*, allowing middleware-controlled data sharing without constraints associated with traditional user accounts, file systems, and administrative domains.

The technique presented in this paper builds on an existing, de-facto standard available for heterogeneous platforms — the Network File System, NFS. The virtual file system is realized via extensions to existing NFS implementations that allow reuse of unmodified clients and servers of conventional operating systems: the proposed modifications are encapsulated in a software proxy that is configured and controlled by the computational grid middleware. Therefore, in addition to enabling data sharing, the virtual file system can leverage complementary coordination capabilities from legacy solutions based on NFS.

The described approach is unique in its ability to integrate unmodified applications (even commercial ones) and existing computing infrastructure into a heterogeneous, wide-area network computing environment. This work was conducted in the context of PUNCH [9, 11], a platform for Internet computing that turns the World Wide Web into a distributed computing portal. It is designed to operate in a distributed, limited-trust environment that spans multiple administrative domains. Users can access and run applications via standard Web browsers. Applications can be installed "as is" in as little as thirty minutes. Machines, data, applications, and other computing services can be located at different sites and managed by different entities. PUNCH has been operational for five years — today, it is routinely used by about 2,000 users from two dozen countries. It provides access to more than 70 engineering applications from six vendors, sixteen universities, and four research centers.

This paper describes an implementation of a virtual file system based on call-forwarding NFS proxies and discusses its application to data sharing and coordination on computational grids. This paper also presents results from a quantitative analysis of the performance of the virtual file system. Results show that the performance of the virtual file system in a wide-area network setup is within 1% of the local-area setup for a typical PUNCH compute-intensive application (SimpleScalar [3]). However, for an I/O-intensive application (Andrew) the wide-area execution time is 5.5 times larger than the local-area time.

The paper is organized as follows. Section 10.2 describes the core concepts behind logical user accounts and PUNCH's virtual file system. Section 10.3 describes how the virtual file system can be employed to enable file sharing and coordination on a computational grid. Section 10.4 quantitatively analyzes the performance of the virtual file system. Section 10.5 outlines related work and Section 10.6 presents concluding remarks.

10.2 VIRTUAL FILE SYSTEMS: DECOUPLING DATA MANAGEMENT FROM LOCATION

Today's computing systems tightly couple users, data, and applications to the underlying hardware and administrative domain. For example, users are tied to individual machines by way of user accounts, while data and applications are typically tied to a given administrative domain by way of a local file system. This poses constraints to the allocation of both computation and storage servers in the context of large computational grids [10].

In order to deliver computing as a service in a scalable manner, it is necessary to effect a fundamental change in the manner in which users, data, and applications are associated with computing systems and administrative domains. This change can be brought about by introducing an abstraction layer consisting of 1) logical user accounts [10], and 2) a virtual file system [5]. A network operating system, in conjunction with an appropriate resource management system, can then use these components to build systems of systems at run-time.

The components of a logical account are traditional system accounts that are divided into two categories according to their functionality: *shadow accounts*, which can be dynamically allocated during a computing session, and *file accounts*, which store user files and directories [10].

10.2.1 The PUNCH Virtual File System

A *virtual file system* establishes a dynamic mapping between a user's data residing in a file account and the shadow account that has been allocated for that user. It also guarantees that any given user will only be able to access files that he/she is authorized to access. There are different ways in which such virtual file system functionality can be achieved. In the context of PUNCH, several alternatives have been investigated; this section describes the solution that is currently in use by PUNCH [5].

The PUNCH Virtual File System — PVFS — uses call-forwarding proxies that are configured and controlled by grid middleware. The PVFS proxies interface with *unmodified* NFS clients and servers of existing operating systems through a remote procedure call (RPC) layer. The functionality encapsulated in the proxy allows the middleware to control access to data across logical accounts at the granularity of a single PUNCH user. This proxy-based approach is attractive for two reasons: it works with standard O/S services, and it is relatively simple to implement — the proxies only need to receive, modify and forward standard RPC calls, and need not implement neither client nor server functionality.

153

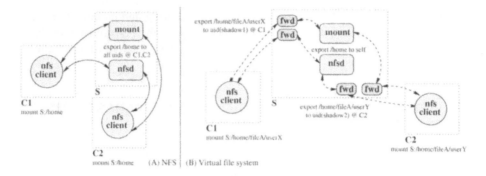

Fig. 10.1 *Overview of conventional (A) and virtual (B) file systems. There are two clients (C1, C2) and one server (S). In (A), the NFS clients C1, C2 share a single logical server via a static mount point for all users under /home. In (B), the file account resides in /home/fileA, and two grid users (X, Y) access the file system through shadow accounts 1 and 2, respectively. The virtual file system clients connect to two independent (logical) servers and have dynamic mount points for users inside /home/fileA that are valid only for the duration of a computing session.*

Although it is based on a standard protocol, the virtual file system approach differs fundamentally from traditional file systems. For example, with NFS, a file system is established once on behalf of multiple users by system administrators (Figure 10.1A). In contrast, the virtual file system creates and terminates *dynamic* client-server sessions that are managed on a *per-user* basis by the grid middleware; each session is only accessible by a given user from a specified client, and that too only for the duration of the computing session (Figure 10.1B). The following discussion outlines the sequence of steps involved in the setup of a PVFS session.

When a user attempts to initiate a run (or session), a compute server and a shadow account (on the compute server) are allocated for the user by PUNCH's *active yellow pages* service [16]. Next, the file service manager spawns a proxy daemon in the file account of the server in which the user's files are stored. This daemon is configured to only accept requests from one user (Unix uid of shadow account) on a given machine (IP address of compute server). Once the daemon is configured, the mount manager employs the standard Unix "mount" command to mount the file system (via the proxy) on the compute server.

Once the PVFS session is established, all NFS requests originating from the compute server by a given user (i.e., shadow account) are processed by the proxy. For valid NFS requests, the proxy modifies the user and group identifiers of the shadow account to the identifiers of the file account in the arguments of NFS remote-procedure calls; it then forwards the requests to the native NFS server. If a request does not match the appropriate user-id and IP credentials of the shadow account, the request is denied and is not forwarded to the native server.

Figure 10.1B shows the client-side mount commands issued by the compute servers "C1" and "C2", under the assumption that the PUNCH file account has user accounts laid out as sub-directories of /home/fileA in file server "S". The path

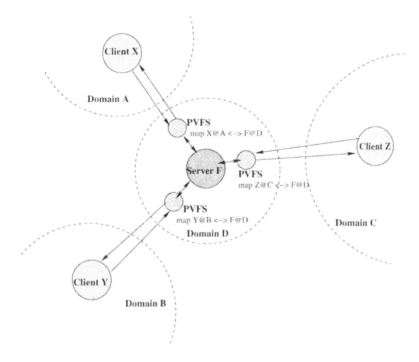

Fig. 10.2 *Sharing data via logical user accounts and PVFS: users with shadow accounts X, Y, Z on clients at domains A, B, C share data from a single file account (F) located in a fourth domain (D).*

exported by the mount proxy ensures that userX cannot access the parent directory of /home/fileA/userX (i.e., this user cannot access files from other users).

10.3 ENABLING FILE SHARING AND COORDINATION

On a typical usage scenario of the virtual file system, a single PVFS proxy enables the dynamic mapping between one shadow account and one file account across administrative domains (Figure 10.1). In a scenario of coordination among several entities, the virtual file system can be setup in a different manner (Figure 10.2):

- Multiple proxies - one per sharer - are spawned on the server side.

- Each proxy authenticates to a single sharer, and maps the identity of the (remote) shadow account of the sharer to the (local) file account of the data server.

The setup of Figure 10.2 allows user-transparent sharing of a single file system among distributed users. On the server side, NFS requests are processed by the native O/S server as if they originated from a *single* user on the *local* machine. In other words, the proxies support multiplexing of the file account by different users. On

155

the client side, the (reverse) mapping of user identities from file to shadow account ensures that each user has appropriate privileges to access the file system.

The grid middleware is responsible for implementing the mechanisms to support data sharing through the virtual file system. Currently, PUNCH does not support sharing mechanisms and only implements a single virtual file system. Nonetheless, a model supporting data sharing towards coordination can be built on top of existing services built for single grid users. The existing virtual file system is built with support of three services [5]: the active yellow pages service, responsible for allocating shadow accounts; the server-side proxy manager, responsible for starting up and shutting down PVFS proxies; and the client-side mount manager, responsible for establishing and closing NFS file system connections. A model supporting data sharing across multiple users can be built by implementing, in the grid middleware, mechanisms to synchronize the issuing of multiple services (e.g. yellow-page queries for all sharers) for one computing session. These mechanisms are necessary only at the beginning and end of a coordination session; during the actual session, the grid middleware is not involved in the file system transactions, and no changes to the PVFS proxy daemon are necessary.

The file system sharing mechanism supported via PVFS, coupled with the support for unmodified applications and the "de-facto" standard network file system semantics of NFS, provide the basic functionality to support coordination across a computational grid. File-based mechanisms to support locking, concurrency control and ordering among distributed processes can be implemented on top of the virtual file system, and also reused from unmodified applications that already provide these mechanisms on native NFS setups.

The performance of the virtual file system in such coordination environment is dependent on the performance of the underlying NFS protocol for data transfers and for synchronization operations. Although file systems specifically designed for wide-area environments (e.g., AFS [14, 17]) may deliver better performance than NFS, they are not commonly available in standard machine configurations, and hence would be difficult to build upon in grids. The following section presents PVFS performance results for single-client data transfers across wide-area networks; the performance of PVFS for synchronization operations across clients is subject of future work.

10.4 PERFORMANCE

10.4.1 NFS versus PVFS

The relative performance of PVFS can be measured with respect to native NFS in terms of its impact on the number of transactions per second. PVFS introduces a fixed amount of overhead for each file system transaction. This overhead is primarily a function of RPC handling and context switching; the actual operations performed by the proxy are very simple and independent of the type of NFS transaction that is forwarded. Previous work has investigated the performance overhead introduced by PVFS with respect to native NFS; results show that (1) the overhead is sensitive to the processing performance of the server, and (2) for a fast server the average overhead is 3% for single-client executions of the Andrew benchmark [10].

10.4.2 Local-area versus Wide-area

The performance of PVFS across wide-area networks has, in it current implementation, an upper bound given by the wide-area performance of NFS[1]. The following performance analysis characterizes the wide-area performance of PVFS for two different types of applications; it is based on the execution of the Andrew file system benchmark (AB [8]) and the SimpleScalar [3] simulator on directories mounted through PVFS. Andrew consists of a sequence of Unix commands of different types (directory creation, file copying, file searching, and compilation) that models a workload typical of a software development environment. SimpleScalar consists of a cycle-accurate simulator of super-scalar microprocessors; it represents a typical compute-intensive engineering application that interfaces with the file system through standard I/O operations.

The experimental setup consists of the set of machines described in Table 10.1. This setup allows the investigation of a wide-area PVFS deployment; results for local-area PVFS deployments have been reported elsewhere [5]. The main objective of the experiment is to characterize the performance of PVFS in an existing grid environment, rather than to investigate a wide range of possible design points.

Machine	#CPUs	CPU type	Memory	Network	O/S
S1	2	933MHz P III	512MB	100Mb/s	Linux 2.4
C1-L	1	900MHz P-III	256MB	100Mb/s	Linux 2.4
C1-W	1	900MHz P-III	256MB	1Mb/s	Linux 2.4

Table 10.1 *Configuration of server (S1) and clients (C1-L, C1-W) used in the performance evaluation. C1-W is connected to a wide-area network through a cable-modem, while C1-L is connected to a local-area switched Ethernet network.*

10.4.3 Wide-area performance

This analysis considers virtual file systems established across wide-area networks. The experimental setup consists of a PVFS client (C1) that connects to a server (S1)

---

[1]Performance enhancements specifically targeted for wide-area applications that are not found in conventional NFS, such as session-customized caching and prefetching, are possible for PVFS; these enhancements are not in place in the current version of PVFS and are subject of future research.

	Client C1-L			
	Mean	**Stdev**	**Min**	**Max**
Andrew	25.7s	3.8s	22s	44s
SimpleScalar	2155s	17s	2136s	2183s
	Client C1-W			
	Mean	**Stdev**	**Min**	**Max**
Andrew	141.4s	7.6s	132s	167s
SimpleScalar	2142s	20s	2136s	2148s

Table 10.2 *Mean, standard deviation, minimum and maximum execution times of 100 samples of Andrew benchmark and 10 samples of SimpleScalar runs on local- and wide-area virtual file systems (clients C1-L and C1-W). The file server is S4. SimpleScalar simulates the "test" dataset of the Spec95 benchmark Tomcatv.*

via two different networks: a switched-ethernet LAN (C1-L) and a residential cable-modem WAN (C1-W, Table 10.1)[2].

Virtual file systems are established on top of native Linux NFS-V2 clients and servers with 8KB read/write transfer sizes. The experiments consider results from executions of Andrew (100 samples) and SimpleScalar (10 samples), collected at intervals of 30 or more minutes to avoid possible cache-induced interference.

Table 10.2 summarizes the performance data from this experiment. The results show that the average wide-area performance of a compute-intensive application under PVFS is very close (within 1%) to the local-area performance. However, for the I/O-intensive application Andrew, the local-area setup delivers a 5.5 speedup relative to wide-area.

The difference in performance can be attributed to a combination of the larger latency and smaller bandwidth of the wide-area setup. About 70% of the RPC requests are latency-sensitive because the amount of data transferred is small (e.g. NFS lookup, getattr) while the remaining 30% are bandwidth-sensitive (e.g. NFS read, write). It is possible to exploit unique characteristics of PVFS to reduce the performance degradation due to smaller bandwidth and longer latency. Directions for future research include the use of middleware-driven techniques for latency-hiding (e.g. prefetching and relaxed consistency models) as well as for improving bandwidth (e.g. aggressive proxy-based caching).

[2]The cable-modem setup delivers around 1Mbit/s download and 100Kbit/s upload bandwidths; actual bandwidths vary depending on the number of network users.

10.5 RELATED WORK

Employing logical user accounts streamlines and distributes many of the typically centralized tasks associated with creating and maintaining user accounts in a distributed computing environment. It also facilitates access control at a finer granularity than is possible with traditional user accounts; the grid-controlled, dynamic multiplexing of (client) shadow accounts onto a single (server) file account allows seamless sharing of a file system across different administrative domains. To our knowledge, PUNCH is the first and, to date, the only system to exploit this mechanism. Related solutions (e.g. Condor [12]) allow execution on remote machines within a *single* user account. However, in order to do this, it requires that applications be relinked with Condor-specific I/O libraries, making this approach unsuitable for situations where object code is not available (e.g. as with commercial applications).

Current grid computing solutions typically employ file staging techniques to transfer files between user accounts in the absence of a common file system. Examples of these include Globus [4] and PBS [7, 2]. As indicated earlier, file staging approaches require the user to explicitly specify the files that need to be transferred, and are often not suitable for session-based or database-type applications. The Kangaroo technique [18] also employs RPC-forwarding agents. However, unlike PVFS, Kangaroo does not provide full support for the file system semantics commonly offered by existing NFS/UNIX deployments (e.g. delete and link operations), and therefore it is not suitable for general-purpose programs that rely on NFS-like file system semantics (e.g. database and CAD applications).

Legion [6, 19] employs a modified NFS daemon to provide a virtual file system.[3] >From an implementation standpoint, this approach is less appealing than call forwarding: the NFS server must be modified and extensively tested for compliance and for reliability. Moreover, current user-level NFS servers (including the one employed by Legion) tend to be based on the older version 2 of the NFS protocol, whereas the call forwarding mechanism described in this paper works with version 3 (the current version).[4] Finally, user-level NFS servers generally do not perform as well as the kernel servers that are deployed with the native operating system.

The Self-certifying File System (SFS) [13] is another example of a virtual file system that builds on NFS. The primary difference between SFS and the virtual file system described here is that SFS uses a single (logical) proxy to handle all file system users. In contrast, PVFS partitions users across multiple, independent proxies. Previous efforts on the Ufo [1] and Jade [15] systems have employed per-user file system agents. However, Jade is not an application-transparent solution — it requires that applications be re-linked to dynamic libraries, while Ufo requires low-level process tracing techniques that are complex to implement and highly O/S-dependent.

---

[3]This approach has also been investigated in the context of PUNCH.
[4]The call forwarding mechanism also works with version 2 of NFS.

10.6 CONCLUSIONS

This paper describes a technique for supporting data sharing and coordination in computational grids, in a manner that is decoupled from the underlying system accounts. A grid environment leverages the computing power of existing networked machines; therefore, grid-oriented solutions must be able to work with standard software solutions that are deployed across its heterogeneous computing nodes. The virtual file system described in this paper is suited for grid deployments because (1) NFS is a "de-facto" standard for network file systems, and (2) PVFS can be deployed on top of existing configurations, with small administrative overheads.

A key property of the virtual file system is the support for per-user file system proxies, which enable middleware-controlled and user-transparent data transfers on top of native NFS. This paper shows that this property can be leveraged to support a model where multiple users share a single account via independent PVFS proxies. The file system sharing mechanism supported via PVFS, coupled with the support for unmodified applications and the "de-facto" standard network file system semantics of NFS, provide the basic functionality to support coordination across a computational grid.

The performance of the virtual file system in such coordination environment is dependent on the performance of the underlying NFS protocol for data transfers and for synchronization operations. This paper shows that the user-perceived wide-area performance of PVFS is good for a compute-intensive application, but 5.5 times worse than the local-area performance for an I/O-intensive application. Future work will investigate its performance for synchronization operations, and wide-area performance enhancements.

Acknowledgements

This work was partially funded by the National Science Foundation under grants EEC-9700762, ECS-9809520, EIA-9872516, and EIA-9975275, and by an academic reinvestment grant from Purdue University. Intel, Purdue, SRC, and HP have provided equipment grants for PUNCH compute-servers.

REFERENCES

1. Albert D. Alexandrov, Maximilian Ibel, Klaus E. Schauser, and Chris J. Scheiman. Ufo: A personal global file system based on user-level extensions to the operating system. *ACM Transactions on Computer Systems*, 16(3):207–233, August 1998.

2. Albeaus Bayucan, Robert L. Henderson, Casimir Lesiak, Bhroam Mann, Tom Proett, and Dave Tweten. Portable Batch System: External reference specification. Technical report, MRJ Technology Solutions, November 1999.

3. Doug Burger and Todd M. Austin. The simplescalar tool set, version 2.0. Technical Report 1342, Computer Sciences Department, University of Wisconsin at Madison, June 1997.

4. Karl Czajkowski, Ian Foster, N. Karonis, Carl Kesselman, Stuart Martin, Warren Smith, and Steven Teucke. A resource management architecture for metacomputing systems. In *Proceedings of the Fourth Workshop on Job Scheduling Strategies for Parallel Processing*, 1998. Held in conjunction with the International Parallel and Distributed Processing Symposium.

5. Renato J. Figueiredo, Nirav H. Kapadia, and José A. B. Fortes. The PUNCH virtual file system: Seamless access to decentralized storage services in a computational grid. In *Proceedings of the 10th IEEE International Symposium on High Performance Distributed Computing (HPDC'01)*, San Francisco, California, August 2001.

6. Andrew S. Grimshaw, William A. Wulf, et al. The Legion vision of a worldwide virtual computer. *Communications of the ACM*, 40(1), January 1997.

7. Robert L. Henderson and Dave Tweten. Portable batch system: Requirement specification. Technical report, NAS Systems Division, NASA Ames Research Center, August 1998.

8. J. H. Howard, M. L. Kazar, S. G. Menees, D. A. Nichols, M. Satyanarayanan, R. N. Sidebotham, and M. J. West. Scale and performance of a distributed file system. *ACM Transactions on Computer Systems*, 6(1):51–81, February 1988.

9. Nirav H. Kapadia, Renato J. O. Figueiredo, and José A. B. Fortes. PUNCH: Web portal for running tools. *IEEE Micro*, pages 38–47, May-June 2000.

10. Nirav H. Kapadia, Renato J. O. Figueiredo, and José A. B. Fortes. Enhancing the scalability and usability of computational grids via logical user accounts and virtual file systems. In *Proceedings of the Heterogeneous Computing Workshop (HCW) at the International Parallel and Distributed Processing Symposium (IPDPS)*, San Francisco, California, April 2001.

11. Nirav H. Kapadia and José A. B. Fortes. PUNCH: An architecture for web-enabled wide-area network-computing. *Cluster Computing: The Journal of Networks, Software Tools and Applications*, 2(2):153–164, September 1999. In special issue on High Performance Distributed Computing.

12. M. Litzkow, M. Livny, and M. W. Mutka. Condor - a hunter of idle workstations. In *Proceedings of the 8th International Conference on Distributed Computing Systems*, pages 104–111, June 1988.

13. David Mazières, Michael Kaminsky, M. Frans Kaashoek, and Emmett Witchel. Separating key management from file system security. In *Proceedings of the 17th ACM Symposium on Operating Systems Principles (SOSP)*, Kiawah Island, South Carolina, December 1999.

14. J. H. Morris, M. Satyanarayanan, M. H. Conner, J. H. Howard, D. S. Rosenthal, and F. D. Smith. Andrew: A distributed personal computing environment. *Communications of the ACM*, 29(3):184–201, March 1986.

15. Herman C. Rao and Larry L. Peterson. Accessing files on the internet: the jade file system. *IEEE Transactions on Software Engineering*, 19(6):613–625, June 1993.

16. Dolors Royo, Nirav H. Kapadia, José A. B. Fortes, and Luis Diaz de Cerio. Active yellow pages: A pipelined resource management architecture for wide-area network computing. In *Proceedings of the 10th IEEE International Symposium on High Performance Distributed Computing (HPDC'01)*, San Francisco, California, August 2001.

17. A. Z. Spector and M. L. Kazar. Wide area file service and the AFS experimental system. *Unix Review*, 7(3), March 1989.

18. Douglas Thain, Jim Basney, Se-Chang Son, and Miron Livny. The kangaroo approach to data movement on the grid. In *Proceedings of the 2001 IEEE International Conference on High-Performance Distributed Computing (HPDC)*, pages 325–333, Aug. 2001.

19. Brian S. White, Andrew S. Grimshaw, and Anh Nguyen-Tuong. Grid-based file access: The Legion I/O model. In *Proceedings of the 9th IEEE International Symposium on High Performance Distributed Computing (HPDC'00)*, pages 165–173, Pittsburgh, Pennsylvania, August 2000.

Author(s) affiliation:

- **Renato J. Figueiredo**

  Department of Electrical and Computer Engineering
  Northwestern University
  Evanston, IL 60208, USA
  Email: renato@ece.nwu.edu

- **Nirav Kapadia**

  Cantiga Systems Inc.
  Mountain View, CA 94043, USA

- **José A. B. Fortes**

  Department of Electrical and Computer Engineering
  University of Florida
  Gainesville, FL 32611, USA
  Email: fortes@ufl.edu

Part III - Intelligent Coordination

and Ubiquitous Computing

11

Terraforming Cyberspace

Jeffrey M. Bradshaw
Niranjan Suri
Maggie Breedy
Alberto Cañas
Robert Davis
Kenneth Ford
Robert Hoffman
Renia Jeffers
Shri Kulkarni
James Lott
Thomas Reichherzer
Andrzej Uszok

11.1 INTRODUCTION

During the 1940s, under the pseudonym of Will Stewart, Jack Williamson published a series of fictional stories describing a process for attaching atmospheres to planets in order to make them capable of sustaining life. 'Terraforming', the term he coined for this activity was first picked up by other science fiction writers. Eventually, it captured the imagination of a small but zealous core of scientists, space advocacy groups, and segments of the public who began focusing on Mars as the most likely target for transformation and eventual colonization. The May 1991 issue of Life Magazine ran a cover story describing a 150-year plan for a Martian metamorphosis through orbiting solar reflectors that would melt polar water, surface factories that would produce needed gases in the atmosphere, and the ultimate planting of hearty plant species as the temperature approached the freezing point of water (figure 11.1). Today many articles, books, and Web sites continue to develop the theme.

Fig. 11.1 Terraforming Mars.

Like pre-terraformed Mars for humans, cyberspace is currently a lonely, dangerous, and relatively impoverished place for software agents (figure 11.2). Though promoted as collaborative, agents do not easily sustain rich long-term peer-to-peer relationships, let alone any semblance of meaningful community involvement. While their features for secure reliable interaction are often touted, there is no social safety net to help agents out when they get stuck, or worse yet to prevent them from setting the network on fire when they go off the deep end. Despite the fact that agent designers want them to communicate at an "almost human" level, agents are cut off from most of the world in which humans operate. Though capable of self-directed mobility, they are hobbled by severe practical restrictions on when and where they can go. Ostensibly endowed with autonomy, an agent's very existence can be terminated unceremoniously by the first passerby who happens to find the power switch.

In consequence of these (and other) limitations, most of today's agents are designed for "solitary, poor, nasty, brutish, and short" lives of narrow purpose in a relatively bounded and static computational world [1]. With rare exception, today's

---

[1] Thomas Hobbes' entire passage is worth reproducing here: "Of the natural condition of Mankind, as concerning their felicity, and misery, whatsoever therefore is consequent to a time of war, where every man is enemy to every man; the same is consequent to the time, wherein men live without other security, than what their own strength, and their own invention shall furnish them withal. In such condition, there is no place for industry, because the fruit thereof is uncertain; and consequently no culture of the earth, no navigation, nor use of the commodities that may be imported by sea; no commodious building, no instruments of moving and removing such things as require much force; no knowledge of the face of the earth, no account of time, no arts, no letters, no society; and which is worst of all, continual fear and danger of violent death; and the life of people, solitary, poor, nasty, brutish, and short" (Leviathan, i . xiii. 9).

Fig. 11.2 Agent life on the wire is "solitary, poor, nasty, brutish, and short." (Brooklyn Bridge workers, 1914, courtesy Collections of the Municipal Archives of the City of New York).

agents are not deployed in critical, long-lived, secure, or high-risk tasks, or on missions requiring widespread agent migration, or the collaboration of large numbers of agents interacting in complex, unpredictable ways.

Progress in some of these limitations necessarily awaits the results of ongoing research in traditional approaches to agent autonomy, collaborativity, adaptivity, and mobility. However, we argue that focusing greater attention not only on making agents smarter and stronger but also on making the environment in which they operate more capable of sustaining various forms of agent life and civilization would simplify some of these problems. A modest terraforming effort would enable not only intelligent agents but also the agent-equilvalent of dogs, insects, and chickens to survive and thrive in cyberspace.

Fortunately, the basic infrastructure with which we can begin the terraforming effort is becoming more available. Designed from the ground up to exploit next-generation Internet capabilities, grid-based approaches aim to provide a universal source of dynamically pluggable, pervasive, and dependable computing power, while guaranteeing levels of security and quality of service that will make new classes of applications possible [14]. By the time these approaches become mainstream for

Does Hobbes' general argument for the institution of government as a check on the self-serving tendency of individuals find a more natural application to artificial agents than to humans?

large-scale applications, they will also have migrated to ad hoc local networks of very small devices [16].

The CoABS Grid, based on Sun's Jini services and developed at Global InfoTek (GITI) under DARPA's Control of Agent-Based Systems (CoABS) program, arguably provides the most successful and widely used infrastructure to date for the large-scale integration of heterogeneous agent frameworks with object-based applications, and legacy systems [9; 21]. Over the next few years, we expect a confluence of this effort with those of the larger computational grid community (http://www.gridforum.org). The Java Agent Services Expert Group (JAS, JSR 87), under the auspices of Sun's Java Community Process (http://www.java-agent.org), the OMG Agent PSIG (http://www.objs.com /agent/), and the FIPA Abstract Architecture Working Group (http://www.fipa.org /activities /architecture.html) are similarly at work on essential contributions to interoperable agent infrastructure.

However, we must go far beyond these current efforts to enable the vision of terraforming cyberspace (Figure 11.3). Current infrastructure implementations typically provide few resource guarantees and no incentives for agents and other components to look beyond their own selfish interests. At a minimum, future infrastructure must go beyond the *bare essentials* to provide pervasive *life support services* (relying on mechanisms such as orthogonal persistence and strong mobility [30; 31]) that help ensure the survival of agents that are designed to live for many years. Beyond the basics of individual agent protection, long-lived agent communities will depend on *legal services*, based on explicit policies, to ensure their rights and help them fulfill their obligations. Benevolent *social services* will also eventually be provided to offer proactive help when needed. Although some of these elements of terraforming for agents exist in embryo within specific agent systems, their scope and effectiveness has been limited by the lack of underlying support at the platform level.

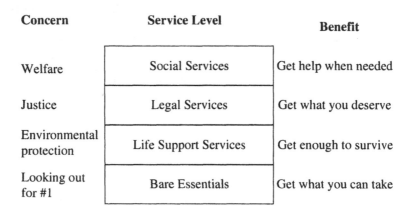

Concern	Service Level	Benefit
Welfare	Social Services	Get help when needed
Justice	Legal Services	Get what you deserve
Environmental protection	Life Support Services	Get enough to survive
Looking out for #1	Bare Essentials	Get what you can take

Fig. 11.3 Elements of terraforming for software agents.

11.2 NOMADS LIFE SUPPORT SERVICES

NOMADS is the name we have given to the combination of Aroma, an enhanced Java-compatible Virtual Machine (VM), with its Oasis agent execution environment [30; 31]. In its current version, it is designed to provide basic *life support services* ensuring agent environmental protection of two kinds:

- assurance of availability of system resources, even in the face of buggy agents or denial-of- service attacks;

- protection of agent execution state, even in the face of unanticipated system failure.

Our approach for life support services has thus far been two-pronged: enabling as much protection as possible in standard Java VMs while also providing NOMADS and the enhanced Aroma VM for those agent applications that require it.

For agents running in standard Java VMs, we create software-based Guards, which enforce policies by relying on the capabilities of the Java 2 security model (including permissions and privileged code wrappers) and the Java Authentication and Authorization Service (JAAS). In contrast to other implementations of Java security, our enhanced JAAS-based approach allow revocation of access permissions under many circumstances as well as the granting of different permissions to different instances of agents from the same code base. For policies that go beyond simple access-based permissions (e.g., obligations, registration policies, conversation policies), Guards implement additional auxiliary KAoS management and enforcement capabilities as required.

Unfortunately, some kinds of protection cannot be provided by merely bolting on new services on top of standard Java VMs. Although Java is currently the most popular and arguably the most mobility-minded and security-conscious mainstream language for agent development, current versions fail to address many of the unique challenges posed by agent software. While few if any requirements for Java mobility, security, and resource management are entirely unique to agent software, typical approaches used in non-agent software are usually hard-coded and do not allow the degree of on-demand responsiveness, configurability, extensibility, and fine-grained control required by agent-based systems.

For agents running in the Aroma VM, we can create a guarded environment that is considerably more powerful in that it not only provides the standard Java and KAoS enforcement capabilities described above, but also supports access revocation under all circumstances, dynamic resource control and full state capture on demand for any Java agent or service.

Protection of agent resources. To fully appreciate the resource control features of Aroma and NOMADS, some understanding of the current Java security model is needed. Early versions of Java relied on the sandbox model to protect mobile code from accessing dangerous methods. In contrast, the security model in the current Java 2 release is *permission-based*. Unlike the original "all or nothing" approach, Java applets and applications can be given varying amounts of access to system resources. Unfortunately, current Java mechanisms do not address the problem of resource control. For example, while it may be possible to prevent a Java program from writing to any directory except /tmp (an access control issue), once the program is given permission to write to the /tmp directory, no further restrictions are placed on the program's I/O (a resource control issue). As another example, there is no way in the current Java implementation to limit the amount of disk space the program may use or to control the rate at which the program is allowed to read and write from the network.

Resource control is important for several reasons. First, without resource control, systems and networks are open to denial of service attacks through resource overuse. Second, resource control lays the foundation for quality-of-service guarantees. Before any quality-of-service guarantees can be made about the availability of resources, the system must be able to limit resource utilization of other tasks (which is currently not possible in the Java environment). Third, resource control presupposes resource accounting, which allows the resources consumed by some component of a system (or the overall system) to be measured for either billing or monitoring purposes. Monitoring resource utilization over time allows the detection of abnormal behavior as part of the system.

Finally, the availability of resource control mechanisms in the environment simplifies the task of developing systems for resource-constrained situations. Consider the task of developing and deploying a new system requiring concurrent execution and resource sharing with existing systems. In such scenarios, the developer of the new system often has to limit the resource utilization of the new system in order to not interfere with the operations of the existing systems (for example, maybe the new system can only use 500 Kb/sec of network bandwidth because the rest of the available network bandwidth is required by the existing systems). Providing such a guarantee requires significant effort on behalf of the developer of the new system. However, if the underlying environment were to provide resource control mechanisms, then the new system could simply make a request to the underlying environment, which can then provide the necessary guarantees.

Aroma currently provides a comprehensive set of resource controls for CPU, disk, and network (Figure 11.4). The resource control mechanisms allow limits to be placed on both the rate and quantity of resources used by Java threads. Rate limits include CPU usage, disk read rate, disk write rate, network read rate and network write rate. Rate limits for I/O are specified in bytes/millisecond. Quantity limits include disk space, total bytes written to disk, total bytes read from the disk, total bytes written to the network, and total bytes read from the network. Quantity limits are specified in bytes. One of the major benefits of the Aroma VM is that resource controls are transparent to the Java code executing inside the VM. In particular, the enforcement of the resource limits does not require any modifications to the Java code. Also, the existence of rate limits (and their enforcement) is completely transparent to the Java component or service.

CPU resource control was designed to support two alternative means of expressing the resource limits. The first alternative is to express the limit in terms of bytecodes executed per millisecond. The advantage of expressing a limit in terms of bytecodes per unit time is that given the processing requirements of a thread, the thread's execution time (or time to complete a task) may be predicted. Another advantage of expressing limits in terms of bytecodes per unit time is that the limit is system and architecture independent. The second alternative is to express the limit in terms of some percentage of CPU time, expressed as a number between 0 and 100. Expressing limits as a percentage of overall CPU time on a host provides better control over resource consumption on that particular host.

Rate limits for disk and network are expressed in terms of bytes read or written per millisecond. If a rate limit is in effect, then I/O operations are transparently delayed if necessary until such time that allowing the operation would not exceed the limit.

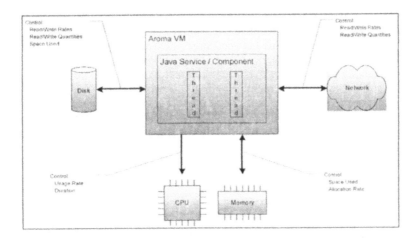

Fig. 11.4 Resource control in NOMADS.

Threads performing I/O operations will not be aware of any resource limits in place unless they choose to query the VM.

Quantity limits for disk and network are expressed in terms of bytes. If a quantity limit is in effect, then the VM throws an exception when a thread requests an I/O operation that would result in the limit being exceeded.

In agent environments, several uses of the NOMADS-based resource control mechanisms are possible. First, the KAoS Domain Manager (explained below) and the VM-level Guard will be able to utilize the resource control capabilities in order to place limits on the resources consumed by services and components running within the Aroma VM. The Guard will be able to vary the resource limits to accommodate changes in policy or level of service guarantees. The Guard will also be able to take advantage of the resource accounting capabilities to measure and report back on the resources consumed by services and components and, if policy permits, to look for patterns of resource abuse that might signal denial-of-service attacks and take autonomous action to reduce resources to the attacker accordingly.

We are working with Sun Microsystems Laboratories to help incorporate resource control capabilities into commercial Java Virtual Machines. Incorporating Aroma-like resource control mechanisms into Java will enable Agent systems and a wide-range of other applications to run in more secure environments.

Protection of agent state. With respect to protection of agent state, we need a way to save the entire state of the running agent or component, including its execution stack, anytime so it can be fully restored in case of system failure or a need to temporarily suspend its execution. The standard term describing this process is checkpointing. Over the last few years, the more general concept of transparent persistence (sometimes called "orthogonal persistence") has also been developed by researchers at Sun Microsystems and elsewhere [20]. The goal of this research is to define language-independent principles and language-specific mechanisms by which persistence can be made available for all data, irrespective of type. Ideally, the approach would not

171

require any special work by the programmer (e.g., implementing serialization methods in Java or using transaction interfaces in conjunction with object databases), and there would be no distinction made between short-lived and long-lived data.

The Aroma VM has been enhanced with the capability to capture the execution state of any running Java program. Current commercial Java VMs do not provide any mechanisms to capture the execution state of a Java program. The state capture mechanism is used to provide strong mobility for NOMADS agents, which allows them to request mobility no matter at what point they are in running their code. Without strong mobility, the code for a mobile agent needs to be structured in a special way to accommodate migration operations. Strong mobility allows agents to be mobile without any special structural requirements (see discussion of resource redirection below).

Promising work has been done on translating agents that use strong mobility into agents that use weak mobility xADDINENRfu. These approaches work well for agents that are single threaded and do not require asynchronous state capture. By asynchronous, we mean a request to capture state that is generated by an external unexpected event or interrupt. The Aroma VM supports capturing of execution state of both multi-threaded agents and allows external events to trigger state capture operations.

We have used the state capture features of NOMADS extensively for agents requiring anytime mobility, whether in the performance of some task or for immediate escape from a host under attack or about to go down (we call this scenario "scram"). We have also put these features to use for transparent load-balancing and forced code migration on demand in distributed computing applications [32]. To support transparent persistence for agents and agent infrastructure components, we are implementing scheduled and on-demand checkpointing services that will protect agent execution state, even in the face of unanticipated system failure.

Forced migration of agents would fail if the agent were using local resources (such as files or network endpoints) on the original host. To solve this problem, we have also implemented transparent redirection of resource access for files and TCP sockets. For example, network redirection is provided through a mechanism called Mockets (mobile sockets) [26], which allow an agent to keep a TCP socket connection open while moving from host to host. Resource redirection is an important requirement to achieve full forced migration of agents.

11.3 KAOS LEGAL AND SOCIAL SERVICES

Terraforming cyberspace involves more than regulation of computing resources and protection of agent state. As the scale and sophistication of agents grow, and their lifespan becomes longer, agent developers and users will want the ability to express complex high-level constraints on agent behavior within a given environment. It seems inevitable that productive interaction between agents in long-lived communities will also require some kind of legal services, based on explicit enforceable policies, to ensure their rights and help them fulfill their obligations. Over time, it seems likely that benevolent social services will also eventually evolve to offer help with individual agent or systemic problems.

In both legal and social services arenas, it is clear that preventive initiatives are nearly always superior to after-the-fact remedies, as the following verse by Joseph Malins illustrates:

'Twas a dangerous cliff, as they freely confessed,

Though to walk near its crest was so pleasant;

But over its terrible edge there had slipped A duke and full many a peasant.

So the people said something would have to be done,

But their project did not at all tally;

Some said, "Put a fence around the edge of the cliff,"

Some, "An ambulance down in the valley."

We are basing our approach on the assumption that preventive policy-based 'fences' can complement and enhance after-the-fact remedial 'ambulance in the valley' mechanisms. The policies governing some set of agents aim to describe expected behavior in sufficient detail that deviations can be easily anticipated or detected. At the same time, related policy support services help make compliance as easy as possible. Complementing these policy support services, various enforcement mechanisms operate as a sort of 'cop at the top of the cliff' to warn of potential problems before they occur. When, despite all precautions, an accident happens remedial services are called as a last resort to help repair the damage. In this manner, the policy-based fences and the after-the-fact ambulances work together to ensure a safer environment for individual agents and the communities in which they operate.

Overview of policy-based agent management. Policy-based management approaches have grown considerably in popularity over the last few years. Unlike previous versions, the Java 2 security model defines security policies as distinct from implementation mechanism. Access to resources is controlled by a Security Manager, which relies on a security policy object to dictate whether class X has permission to access system resource Y. The policies themselves are expressed in a persistent format such as text so they can be viewed and edited by any tools that support the policy syntax specification. This approach allows policies to be configurable, and relatively more flexible, fine-grained, and extensible. Developers of applications no longer have to subclass the Security Manager and hard-code the application's policies into the subclass. Programs can make use of the policy file and the extensible permission object to build an application whose security policy can change without requiring changes in source code.

The basic policytool Java currently provides, assists users in editing policy files. However, to be useful and usable in realistic settings, policy-based administration tools should contain domain knowledge and conceptual abstractions to allow applications designers to focus their attention more on high-level policy intent than on the details of implementation. Moreover, while Java provides only for static policies, critical agent applications will require tools for the monitoring, visualization, and dynamic modification of policies at runtime.

The scope of policy-based agent management includes typical security concerns such as *authorization, encryption, access* and *resource control* policies, but also goes beyond these in significant ways. For example, KAoS pioneered the concept of agent conversation policies [5; 17]. Teams of agents can be formed, maintained, and disbanded through the process of agent-to-agent communication using an appropriate semantics [8; 10; 33]. Conversation policies assure coherence in the adoption and discharge of team commitments by heterogeneous agents of different levels of sophistication [5; 6]. These conversation policies are designed to assure robust behavior and to keep computational overhead for team maintenance to an absolute minimum

173

[17; 19; 29]. In addition to conversation policies, we are in the process of developing representations and enforcement mechanisms for *mobility policies [23]*, *domain registration policies*, and various forms of *obligation policies* (see below).

There are some important differences between the objectives of our approach and that of others working to encourage and enforce security, robustness, and cooperativity constraints among communities of agents. First, unlike most multi-agent coordination environments, the approach does not assume that we are dealing with a homogeneous set of agents written within the same agent framework. With respect to the environmental protection, legal, and social services functions provided, we aim insofar as possible to put KAoS and non-KAoS agents on the same footing-with little or no modification to the agents themselves required. In fact, because our services aim to protect against the negative effects of buggy or malicious agents, we have to make sure that the policy-management mechanisms are designed to work even when agents are trying to work against them. Second, insofar as possible the framework needs to support dynamic runtime policy changes, and not merely static configurations determined in advance. Third, the framework needs to be extensible to a variety of execution platforms with different enforcement mechanisms-initially Java and Aroma-but in principle any platform for which a Guard may be written. Fourth, the framework must be robust in continuing to manage and enforce policy in the face of attack or failure of any combination of components. Finally, we recognize the need for easy-to-use policy-based administration tools capable of containing domain knowledge and conceptual abstractions that let application designers focus their attention more on high-level policy intent than on implementation details. Such tools require powerful graphical user interfaces for monitoring, visualizing, and dynamically modifying policies at runtime.

In short, the policy management framework must ensure maximum freedom and heterogeneity of the agents and non-intrusiveness of the enforcement mechanisms, while respecting the bounds of human-determined constraints designed to ensure selective conformity of behavior.

DAML-based policy representation. In principle, developers could use a variety of representations to express policies. At one extreme, they might write these policies in some propositional or constraint representation. At the other extreme lie a wide variety of simpler schemes, each of which gives up some types of expressivity. Several considerations affect the choice of representation for a particular application, including composability, computability, efficiency, expressivity, and amenability to various sorts of analysis and inference.

With funding from the DARPA CoABS program, we have developed KAoS Policy Ontologies (KPO). These ontologies are expressed in DAML (http://www.daml.org) and work in conjunction with a set of KAoS policy-management services.

Designed to support the emerging "Semantic Web," DAML is the latest in a succession of Web markup languages [2]. HTML, the first Web markup language, allowed users to markup documents with a fixed set of formatting tags for human use and readability. XML allows users to add arbitrary structures to their documents but expresses very little directly about what the structures mean. RDF (Resource Description Format) encodes meaning in sets of subject-verb-object triples, where elements of these triples may each be identified by a URI (typically a URL).

DAML extends RDF to allow users to specify ontologies composed of taxonomies of classes and inference rules. These ontologies can be used by people for a variety of purposes, such as enabling more accurate or complex Web searches. Agents can also use semantic markup languages to understand and manipulate Web content in

174

significant ways; to discover, communicate, and cooperate with other agents and services; or, as we outline in this paper, to interact with policy-based management and control mechanisms.

The current KPO specification defines basic ontologies for actors, actions, entities that are the targets of actions (e.g., computing resources), places, policies, and policy conditions. We have extended these ontologies to represent simple atomic Java permissions, as well as more complex NOMADS, and KAoS policy constructs. It is expected that for a given application, the ontologies will be further extended with additional classes, individuals, and rules. Individual policies will be put into force as required. Through various property restrictions, a given policy can be variously scoped, for example, either to individual agents, to agents of a given class, to agents belonging to an intensionally- or extensionally-defined groups (e.g., a domain or team), or to agents running in a given physical place or computational environment (e.g., VM).

The actor ontology distinguishes between agents (that generally can only perform *ordinary actions*) and Domain Managers, Guards, and authorized human users, who may variously be permitted or obligated to perform certain *policy actions*, such as approval and enforcement. The policy ontology distinguishes between *authorizations* (i.e., constraints that permit or forbid some action) and *obligations* (i.e., constraints that require some action to be performed, or else serve to waive such a requirement) [12].

The KAoS Policy Ontologies are intended for a variety of purposes. One obvious application is during inference relating to various forms of online or offline analysis. For example, changes or additions to policies in force, or a change in status of an actor (e.g., an agent joining a new domain or moving to a new host) require logical inference to determine first of all which policies are in conflict and second how to resolve these conflicts [24]. We have implemented a general-purpose algorithm for policy conflict detection and harmonization whose current results promise a surprising degree of efficiency and scalability [2]. The ontologies may also be used in policy disclosure management (see below), reasoning about future actions based on knowledge of policies in force, and in assisting users of policy specification tools to understand the implications of defining new policies given the current context and the set of policies already in force.

KAoS policy management architecture. Figure 11.5 shows the major components of the KAoS policy management architecture.

The KAoS Policy Administration Tool (KPAT), a graphical user interface to policy management functionality, has been developed to make policy specification, revision, and application easier for administrators without specialized training. Using KPAT, an authorized user may make changes over the Web to agent policy. Alternatively, trusted components such as Guards may, if authorized, propose policy changes autonomously based on their observation of system events.

Groups of agents are structured into agent domains and subdomains to facilitate policy administration. A given domain can extend across host boundaries and, conversely, multiple domains can exist concurrently on the same host. Depending on whether policy allows, agents may become members of more than one domain at a time.

---

[2] A detailed description our policy conflict detection and resolution process is currently being prepared for publication.

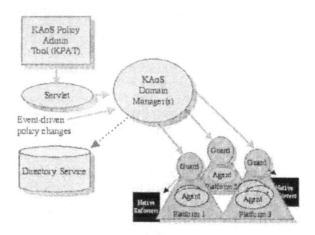

Fig. 11.5 Major components of the KAoS policy management architecture.

KAoS Domain Managers (DM) act in the role of policy decision points to determine whether agents can join their domain and for policy conflict resolution [3]. The DM is responsible for ensuring policy consistency at all levels of a domain hierarchy, for notifying Guards about changes in policy or other aspects of system state that may affect their operation, and for storing state in the directory service. Because DM's are stateless, one DM instance may serve multiple domains or conversely, a single large domain may require several instances of the DM to achieve scalable performance.

Policies are stored within ontologies in the directory service (DS). Although DM's normally provide the limited public interface to the DS, private interfaces may allow the DS to be accessed by other authorized entities in accordance with policy disclosure strategies [27]. For example, trusted agents may be allowed to perform queries concerning domain policies in advance of submitting a registration request to a new domain. Because the policies in the directory service are expressed declaratively, some forms of analysis and verification can be performed in advance and offline, permitting execution mechanisms to be as efficient as possible.

Guards interpret policies that have been approved by the DM and enforce them with appropriate native enforcement mechanisms. While KPAT and the DM, and

---

[3] In the current implementation, DM's delegate inference about policy decisions to the directory service, which incorporates a DAML-based inference engine.

the Guards are intended to work identically across different agent platforms (e.g., DARPA CoABS Grid, Cougaar, Objectspace Voyager) and execution environments (e.g., Java VM, Aroma VM), enforcement mechanisms are necessarily designed for a specific platform and execution environment. Our approach enables policy uniformity in domains that might be simultaneously distributed across multiple platforms and execution environments, as long as semantically equivalent monitoring and enforcement mechanisms are available. Under these conditions, it follows that behavior of agents written using different platforms and running in different execution environments can be kept consistent through the use of these policy-based mechanisms. Because policy analysis and policy conflict resolution normally take place prior to the policy being given to the Guard for enforcement, the operation of the Guards and enforcement mechanisms can be lightweight and efficient.

In applications to date, we have relied on several different kinds of enforcement mechanisms. Enforcement mechanisms built into the execution environment (e.g., OS or Virtual Machine level protection) are the most powerful sort, as they can generally be used to assure policy compliance for any agent or program running in that environment, regardless of how that agent or program was written. For example, the Java Authentication and Authorization Service (JAAS) provides methods that ties access control to authentication. In KAoS, we have developed methods based on JAAS that will allow policies to be scoped to individual agent instances rather than just to Java classes. Currently, JAAS can be used with Java VMs; in the future it should be possible to use JAAS with the Aroma VM as well. As described above, the Aroma VM provides, in addition to Java VM protections, a comprehensive set of resource controls for CPU, disk and network. The resource control mechanisms allow limits to be placed on both the rate and the quantity of resources used by Java threads. Guards running on the Aroma VM can use the resource control mechanisms to provide enhanced security (e.g., prevent or disable denial-of-service attacks), maintain quality of service for given agents, or give priority to important tasks.

A second kind of enforcement mechanism takes the form of extensions to particular agent platform capabilities. Agents that participate in that platform are generally given more permissions to the degree they are able to make small adaptations in their agents to comply with policy requirements. For example, in applications using the DARPA CoABS Grid, we have defined a KAoSAgentRegistrationHelper to replace the default GridAgentRegistrationHelper. Grid agent developers need only replace the class reference in their code to participate in agent domains and be transparently and reliably governed by policies currently in force. On the other hand, agents that use the default GridAgentRegistrationHelper do not participate in domains and as a result they are typically granted very limited permissions in their interactions with domain-enabled agents.

Finally, a third type of enforcement mechanism is necessary for obligation policies. Because obligations cannot be enforced through preventive mechanisms, enforcers can only monitor agent behavior and determine after-the-fact whether a policy has been followed. For example, if an agent is required by policy to report its status every 5 minutes, an enforcer might be deployed to watch whether this is in fact happens, and if not to either try to diagnose and fix the problem, or alternatively take appropriate sanctions against the agent (e.g., reduce permissions or publish the observed instance of noncompliance to an agent reputation service).

Applications and benefits of policy-based approach. An example of the application of KAoS, NOMADS, and Java security policies and mechanisms can be found in our work on the DARPA CoABS-sponsored Coalition Operations Experiment (CoAX) (http:// www.aiai.ed.ac.uk /project/ coax/) [1]. CoAX models military coalition operations and implement agent-based systems to mirror coalition structures, policies, and doctrines. The project aims to show that the agent-based computing paradigm offers a promising new approach to dealing with issues such as

- the interoperability of new and legacy systems,

- the implicit nature of coalition policies,

- security, and

- recovery from attack, system failure, or service withdrawal.

KAoS provides mechanisms for overall management of coalition organizational structures represented as domains and policies, while NOMADS provides strong mobility, resource management, and protection from denial-of-service attacks for untrusted agents that run in its environment.

The combination of the use of libraries of pre-analyzed policy sets, separate policy decision and conflict resolution mechanisms, and efficient policy enforcement mechanisms make the use of policy-based administration tools maximally effective and performant. A policy-based approach has the additional advantages of *reusability, efficiency, context sensitivity*, and *verifiability*:

Reusability. Policies encode sets of useful constraints on agent or component behavior, packaging them in a form where they can be easily reused as the occasion requires. By reusing policies when they apply, we reap the lessons learned from previous analysis and experience while saving ourselves the time it would have taken to reinvent them from scratch.

Efficiency. In addition to lightening the application developers' workload, explicit policies can sometimes increase runtime efficiency. For example, to the extent that policy conflict resolution and conversion of policy to a form that can be used by appropriate enforcement mechanisms can take place in advance, overall performance can be increased.

Context-sensitivity. Explicit policy representation improves the ability of agents, components, and platforms to be responsive to changing conditions, and if necessary reason about the implications of the policies which govern their behavior.

Verifiability. By representing policies in an explicitly declarative form instead of burying them in the implementation code, we can better support important types of policy analysis. First-and this is absolutely critical for security policies-we can externally validate that the policies are sufficient for the application's tasks, and we can bring both automated theorem-provers and human expertise to this task. Second, there are methods to ensure that program behavior which follows the policy will also satisfy many of the important properties of reactive systems: liveness, recurrence, safety invariants, and so forth. Finally, with explicit policies governing different types of agent behavior, we can predict how policies may compose.

11.4 CYBERFORMING TERRASPACE

Tomorrow's world will be filled with agents embedded everywhere in the places and things around us. Providing a pervasive web of sensors and effectors, teams of such agents will function as cognitive prostheses-computational systems that leverage and extend human intellectual, perceptual, and collaborative capacities, just as the steam shovel was a sort of muscular prosthesis or the eyeglass a sort of visual prosthesis [13]. Thus the focus of AI research is destined to shift from Artificial Intelligence to Augmented Intelligence [4; 18]. It is clear that terraforming cyberspace is just the first step; the next step will be to cyberform terraspace, giving agents permanent footholds in the material world.

While simple robotic assistants of various kinds today capture our attention, the future surely holds much more interesting and amazing agent-powered devices than we can currently imagine. A key requirement for such devices is for real-time cooperation with people and with other autonomous systems. While these heterogeneous cooperating entities may operate at different levels of sophistication and with dynamically varying degrees of autonomy, they will require some common means of representing and appropriately participating in joint tasks. Just as important, developers of such systems will need tools and methodologies to assure that such systems will work together reliably, even when they are designed independently.

One example that presages-albeit primitively-such developments is the Personal Satellite Assistant (PSA), a softball-sized flying robot that is being designed to operate onboard spacecraft in pressurized micro-gravity environments (figure 11.6) [15]. First proposed and championed by Yuri Gawdiak of NASA Ames, the PSA will incorporate environmental sensors for gas, temperature, and fire detection, providing the ability for the PSA to monitor spacecraft, payload and crew conditions. Video and audio interfaces and speech understanding capabilities will support for navigation, remote monitoring, and video-conferencing. Ducted fans will provide propulsion and batteries will provide portable power. Most importantly, it is intended to work in close interaction with groups of people and artificial agents, which poses daunting challenges to researchers and developers.

With funding and collaboration from NASA Ames, and RIACS, we are investigating issues in human-robotic teamwork and adjustable autonomy for the PSA [7; 8]. Although we currently envision the PSA as the most accessible and practical initial testbed for our prototyping work in the design of collaborative robots, we are confident that our results will generalize to future cooperative autonomous systems of many other sorts. For instance, future human missions to the Moon and to Mars will undoubtedly need the increased capabilities for human-robot collaborations we envision. Astronauts will live, work, and perform laboratory experiments in collaboration with robots not only inside, but also outside the habitat on planetary surfaces.

Our approach to design of cooperative autonomous systems requires first and foremost a thorough understanding of the kinds of interactive contexts in which humans and autonomous systems will cooperate. With our colleagues at RIACS, we have begun to investigate the use of Brahms [28] as an agent-based design toolkit to model and simulate realistic work situations in space. The agent-based simulation in Brahms will eventually become the basis for the design of PSA functions for actual operations. On its part, IHMC is enhancing the KAoS and NOMADS agent frameworks to incorporate explicit general models of teamwork, mobility, and resource control appropriate for space operations scenarios. We expect these models to be mostly represented in the form of policies [7].

The power of general-purpose teamwork models in multi-agent systems usually comes at a high price. Joint intention theory, for example, is built on an extremely powerful logical framework that includes explicit representation of mental attitudes like belief, goal, intention, and so forth [10; 11; 33]. These attitudes are modeled in the traditional way: as new modal operators in a quantified modal logic. Hence,

179

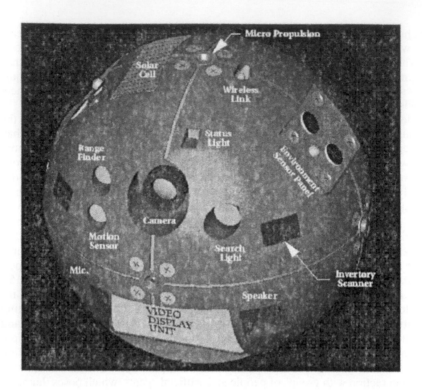

Fig. 11.6 NASA's Personal Satellite Assistant or PSA (Courtesy NASA Ames).

while the most general form of joint intention theory is representationally very attractive, it is often computationally intractable. This tension between expressivity and computability is not limited to teamwork theories; in fact, it is a hallmark of all mentalistic theories of agent behavior and speech-act based agent communication. Thus, when designing agents that include strong teamwork assumptions and powerful communication languages (as do the PSA and other robots), it is critically important to reduce the power of these general models in a way that is sensitive to the agent's domain and expected range of action.

Our approach to building a model of teamwork seeks to incorporate the best of previous research on human-centered collaboration and teamwork, while simultaneously grounding such research in our own work practice study experience. In addition, we are assessing the contributions of allied fields ranging from cognitive function analysis [3], to studies of animal displays, the roles of values and affect, the enablers of effective delegation in humans [25].

By using the analysis and simulation capability in Brahms, we will be able to incorporate models of the PSA work environment and practices in our decisions about how to strategically weaken general joint intention theory without compromising the PSA's ability to perform in its environment. In this way, we will balance empirical analysis, simulation, and top-down theoretical considerations in arriving at a teamwork theory that will allow the PSA to meet the scenario goals. Teams of KAoS-NOMADS-Brahms agents will be formed, maintained, and disbanded through the process of agent-to-agent communication using an appropriate semantics. Agents representing various team members, from humans to autonomous systems to simple devices and sensors, will assure coherence in the adoption and discharge of team commitments and will encapsulate state information associated with each entity. Agent conversation policies will be designed to assure robust behavior and to keep computational overhead for team maintenance to an absolute minimum [6; 17; 19].

11.5 TERRAFORMING TERRASPACE

In discussing requirements for habitable agent environments, we cannot forget that humans have corresponding needs. Terraspace, in many respects, is as badly in need of (re)terraforming as cyberspace [4]. To this end, IHMC is collaborating with the developers of Seaside, Florida, a small, sophisticated Gulf Coast town that is the birthplace of New Urbanism [22](figure 11.7). The New Urbanism movement includes urban designers, environmentalists, transportation experts, social justice advocates, and others who are working together to change American land use from highway-oriented sprawl. The movement proposes reviving and updating traditional town-building principles to produce human-scaled settlements in which numerous pathways connect a variety of buildings internally and to the surrounding landscape. IHMC and the Seaside Institute are hosting a series of workshops, study groups, and visiting scholar programs exploring the application of well-evolved urban design principles to the newly burgeoning and chaotic world of electronic space design.

Increasingly, cyberspace is being perceived in spatial terms for human users. We speak and think of visiting websites, for example, and of the problems of getting lost while trying to navigate through the tangle of hyperlinks, rather than of the packets of information being trucked to our computers on the information superhighway. In any case, very few people live and work on superhighways. As virtual reality technologies are coupled with high bandwidth connectivity, the perception of cyberspace as space will become a reality. Cyberspace will be built to fit the human sense of space.

As a new kind of space for human beings, cyberspace is now woefully primitive. Most of our electronic built space is a rat's nest of bewildering pathways of indeterminate destination, much like medieval Rome. The humanist Popes of the Renaissance used the ideas of the citta ideale to produce connectivity and impart legibility to the layout of their city. In the process, they produced some of the world's most memorable, elegant, and comfortable streets and squares, which continue to provide an environment for all kinds of people engaged in all manner of activities.

Civilization begins when human beings find places to be, make these places their homes, and then create ways to communicate and work together, which depend on their chosen locations. Those who are designing and building cyberspace might benefit from "lessons learned" over thousands of years by those who have been engaged in designing humanity's best physical environments. Moreover, urban designers may yet have something to learn from the architects of cyberspace.

---

[4] In an interesting twist, Jack Williamson, who invented the term "terraforming" to describe an alien world altered for human habitation in his 1942 novel "Seetee Ship," has used the seemingly paradoxical title "Terraforming Earth" as the name of his latest work of science fiction [34]. The book describes the efforts of scientists to bring life back to the earth after asteroids trigger a new Ice Age.

Fig. 11.7 Odessa Street in Seaside, Florida. Courtesy Steven Brooke.

Ultimately both humans and agents have much to gain through the terraforming and cyberforming of the dual realms of atoms and bits. People will feel more welcome and at home in both places. Agents will be freed from their current role as short-lived visitors on the wire to permanent colonists in real and virtual worlds where we can feel comfortable with not knowing or caring exactly where their programs are being physically hosted. They will truly live among us and we will wonder how we ever lived without them.

11.6 ACKNOWLEDGEMENTS

This is an updated and expanded version of an article that originally appeared in IEEE Intelligent Systems, July 2001, pp. 49-56. The authors gratefully acknowledge the support of the DARPA CoABS, Augmented Cognition, and Ultra*Log Programs, the NASA Cross-Enterprise and Intelligent Systems Programs, and the National Technology Alliance, Fujitsu Labs, and The Boeing Company while preparing this paper. Thanks also to many other colleagues who have assisted in this work, including Mark Adler, David Allsopp, Alessandro Acquisti, Patrick Beautement, Todd Carrico, Janet Cerniglia, Bill Clancey, Rob Cranfill, Grzegorz Czajkowski, Chris Dellarocas, Greg Dorais, Mark Greaves, Jack Hansen, Pat Hayes, Jim Hendler, Greg Hill, Heather Holmback, Wayne Jansen, Martha Kahn, Mike Kerstetter, Mike Kirton, Mark Klein, Gerald Knoll, Ora Lassila, Henry Lieberman, Mike Mahan, Framk McCabe, Nicola Muscettola, Debbie Prescott, Anil Raj, Phil Sage, Dylan Schmorrow, Kent Seamons, Maarten Sierhuis, Austin Tate, Al Underbrink, Ron Van Hoof, Doyle Weishar, Alex Wong, Tim Wright, and members of the DARPA CoABS Coalition Operations Experiment (CoAX), the Java Agent Services Expert Group (JSR 87) and the FIPA Abstract Architecture Working Group.

REFERENCES

1. D. Allsopp, P. Beautement, J. M. Bradshaw, E. Durfee, M. Kirton, C. Knoblock, N. Suri, A. Tate, and C. Thompson (2002). Coalition Agents eXperiement (CoAX): Multi-agent cooperation in an international coalition setting. A. Tate, J. Bradshaw, and M. Pechoucek (Eds.), Special issue of IEEE Intelligent Systems.

2. T. Berners-Lee, J. Hendler, and O. Lassila (2001). The semantic web. Scientific American.

3. G. Boy. (1998). Cognitive Function Analysis., Stamford, CT: Ablex Publishing.

4. J. M. Bradshaw, P. Beautement, S. Kulkarni, N. Suri, and A. Raj (2002). Toward a deliberative and reactive agent architecture for augmented cognition. DARPA Augmented Cognition Program White Paper., Pensacola, FL: Institute for Human and Machine Cognition, University of West Florida.

5. J. M. Bradshaw, S. Dutfield, P. Benoit, and J. D. Woolley (1997). KAoS: Toward an industrial-strength generic agent architecture. In J. M. Bradshaw (Ed.), Software Agents. (pp. 375-418). Cambridge, MA: AAAI Press/The MIT Press.

6. J. M. Bradshaw, M. Greaves, H. Holmback, W. Jansen, T. Karygiannis, B. Silverman, N. Suri, and A. Wong (1999). Agents for the masses: Is it possible to make development of sophisticated agents simple enough to be practical? IEEE Intelligent Systems(March-April), 53-63.

7. J. M. Bradshaw, M. Sierhuis, P. Feltovich, A. Acquisti, R. Jeffers, D. Prescott, N. Suri, A. Uszok, and R. Van Hoof (2002). What we can learn about human-agent teamwork from practice. Conference on Autonomous Agents and Multi-Agent Systems (AAMAS 2002). Bologna, Italy, Proceedings of the AAMAS Workshop on Teamwork and Coalition Formation.

8. J. M. Bradshaw, M. Sierhuis, Y. Gawdiak, R. Jeffers, N. Suri, and M. Greaves (2001). Adjustable autonomy and teamwork for the Personal Satellite Assistant. Proceedings of the IJCAI-01 Workshop on Autonomy, Delegation, and Control: Interacting with Autonomous Agents. Seattle, WA, USA.

9. J. M. Bradshaw, N. Suri, M. Kahn, P. Sage, D. Weishar, and R. Jeffers (2001). Terraforming cyberspace: Toward a policy-based grid infrastructure for secure, scalable, and robust execution of Java-based multi-agent systems. Proceedings of the Second International Workshop on Infrastructure for Agents, Mobile Agent Systems, and Scalable Mobile Agents Systems at the Fifth International Conference on Autonomous Agents (Agents-2001). Montreal, Quebec, Canada.

10. P. R. Cohen, and H. J. Levesque (1991). Teamwork. Technote 504. Menlo Park, CA: SRI International, March.

11. P. R. Cohen, H. R. Levesque, and I. Smith (1997). On team formation. In J. Hintikka and R. Tuomela (Ed.), Contemporary Action Theory. Synthese.

12. N. Damianou, N. Dulay, E. C. Lupu, and M. S. Sloman (2000). Ponder: A Language for Specifying Security and Management Policies for Distributed Systems,

Version 2.3. Imperial College of Science, Technology and Medicine, Department of Computing, 20 October 2000.

13. K. M. Ford, C. Glymour, and P. Hayes (1997). Cognitive prostheses. AI Magazine, 18(3), 104.

14. I. Foster, and C. Kesselman (Ed.). (1999). The Grid: Blueprint for a New Computing Infrastructure. San Francisco, CA: Morgan Kaufmann.

15. Y. Gawdiak, J. M. Bradshaw, B. Williams, and H. Thomas (2000). R2D2 in a softball: The Personal Satellite Assistant. H. Lieberman (Ed.), Proceedings of the ACM Conference on Intelligent User Interfaces (IUI 2000), (pp. 125-128). New Orleans, LA, New York: ACM Press.

16. N. A. Gershenfeld (1999). When Things Start to Think., New York: Henry Holt and Company.

17. M. Greaves, H. Holmback, and J. M. Bradshaw (2001). Agent conversation policies. In J. M. Bradshaw (Ed.), Handbook of Agent Technology. (pp. in preparation). Cambridge, MA: AAAI Press/The MIT Press.

18. S. Hamilton (2001). Thinking outside the box at IHMC. IEEE Computer, 61-71.

19. H. Holmback, M. Greaves, and J. M. Bradshaw (1999). A pragmatic principle for agent communication. J. M. Bradshaw, O. Etzioni, and J. Mueller (Ed.), Proceedings of Autonomous Agents '99, (pp. 368-369). Seattle, WA, New York: ACM Press.

20. M. Jordan, and M. Atkinson (1998). Orthogonal persistence for Java-A mid-term report. Sun Microsystems Laboratories.

21. M. Kahn, and P. Sage (2000). DARPA Control of Agent-Based Systems Grid Tutorial. J. M. Bradshaw (Ed.), PAAM 2000. Manchester, England.

22. P. Katz, and V. Jr. Scully (1993). The New Urbanism: Toward an Architecture of Community., New York, NY: McGraw-Hill.

23. G. Knoll, N. Suri, and J. M. Bradshaw (2001). Path-based security for mobile agents. Proceedings of the First International Workshop onthe Security of Mobile Multi-Agent Systems (SEMAS-2001) at the Fifth International Conference on Autonomous Agents (Agents 2001), (pp. 54-60). Montreal, CA, New York: ACM Press.

24. E. C. Lupu, and M. S. Sloman (1999). Conflicts in policy-based distributed systems management. IEEE Transactions on Software Engineering-Special Issue on Inconsistency Management.

25. A. E. Milewski, and S. H. Lewis (1994). Design of intelligent agent user interfaces: Delegation issues. A&T Corporate Information Technologies Services Advanced Technology Planning, October 20.

26. T. R. Mitrovich, K. M. Ford, and N. Suri (2001). Transparent redirection of network sockets. OOPSLA WorkshBp on Experiences with Autonomous Mobile Objects and Agent-based Systems.

27. K. E. Seamons, M. Winslet, and T. Yu (2001). Limiting the disclosure of access control policies during automated trust negotiation. Proceedings of the Network and Distributed Systems Symposium.

28. M. Sierhuis (2001) Brahms: A Multi-Agent Modeling and Simulation Language for Work System Analysis and Design. Doctoral, University of Amsterdam.

29. I. A. Smith, P. R. Cohen, J. M. Bradshaw, M. Greaves, and H. Holmback (1998). Designing conversation policies using joint intention theory. Proceedings of the Third International Conference on Multi-Agent Systems (ICMAS-98), (pp. 269-276). Paris, France, Los Alamitos, CA: IEEE Computer Society.

30. N. Suri, J. M. Bradshaw, M. R. Breedy, P. T. Groth, G. A. Hill, and R. Jeffers (2000). Strong Mobility and Fine-Grained Resource Control in NOMADS. Proceedings of the 2nd International Symposium on Agents Systems and Applications and the 4th International Symposium on Mobile Agents (ASA/MA 2000). Zurich, Switzerland, Berlin: Springer-Verlag.

31. N. Suri, J. M. Bradshaw, M. R. Breedy, P. T. Groth, G. A. Hill, R. Jeffers, T. R. Mitrovich, B. R. Pouliot, and D. S. Smith (2000). NOMADS: Toward an environment for strong and safe agent mobility. Proceedings of Autonomous Agents 2000. Barcelona, Spain, New York: ACM Press.

32. N. Suri, P. T. Groth, and J. M. Bradshaw (2001). While You're Away: A system for load-balancing and resource sharing based on mobile agents. R. Buyya, G. Mohay, and P. Roe (Ed.), Proceedings of the First IEEE/ACM International Symposium on Cluster Computing and the Grid, (pp. 470-473). Brisbane, Australia, Los Alamitos, CA: IEEE Computer Society.

33. M. Tambe, W. Shen, M. Mataric, D. V. Pynadath, D. Goldberg, P. J. Modi, Z. Qiu, and B. Salemi (1999). Teamwork in cyberspace: Using TEAMCORE to make agents team-ready. Proceedings of the AAAI Spring Symposium on Agents in Cyberspace. Menlo Park, CA, Menlo Park, CA: The AAAI Press.

34. J. Williamson (2001). Terraforming Earth., New York, NY: Tor Books.

Authors affiliations:

- **Jeffrey M. Bradshaw, Niranjan Suri, Maggie Breedy, Alberto Cañas, Robert Davis, Kenneth Ford, Robert Hoffman, Renia Jeffers, Shri Kulkarni, James Lott, Thomas Reichherzer, and Andrzej Uszok**

  Institute for Human and Machine Cognition (IHMC)
  University of West Florida
  40 South Alcaniz St.
  Pensacola, FL 32501, USA
  Email: [jbradshaw, nsuri, mbreedy, acanas, rdavis, kford, rhoffman, rjeffers, skulkarni, jlott, treichhe, auszok]@ai.uwf.edu

12

Towards a Notion of Agent Coordination Context

Andrea Omicini

Abstract

On the one hand, recent studies on the history of human societies suggest that the role of the *environment* has to be taken explicitly into account in order to understand evolution of individuals and groups in any non-trivial setting. On the other hand, the notion of *context* is well-known and relevant to several research areas such as natural language, philosophy, logic, and artificial intelligence. In these areas, contexts are typically used to model the effect of the environment (in its most general sense, including the spatial and temporal interpretation of the term) on the communication occurring amongst active entities, such as humans or artificial agents.

Generalising upon the recently introduced notion of *context-dependent coordination*, in this seminal paper we propose the notion of *agent coordination context* as a means to model and shape the space of agent interaction and communication. From a theoretical perspective, agent coordination contexts can serve the purposes of (i) enabling agents to model the environment where they interact and communicate (the *subjective* viewpoint), and (ii) providing a framework to express how the environment affects interpretation of agent communication acts (the *objective* viewpoint). From an engineering perspective, the notion of agent coordination context enables in principle agents to perceive the space where they act and interact, reason about the effect of their actions and communications, and possibly affect their environment to achieve their goals. Also, agent coordination contexts allow engineers to encapsulate rules for governing applications built as agent systems, mediate the interactions amongst agents and the environment, and possibly affect them so as to change global application behaviour incrementally and dynamically.

12.1 EVOLUTION OF SOCIAL SYSTEMS

12.1.1 A Look Back to Human Societies

The complexity of artificial systems is growing day by day, and their dynamics in the era of the Internet and Ubiquitous Computing [36] is slowly but steadily approaching the dynamics of biological and social systems. So, the first (obvious, but maybe the most relevant) question to be answered is: Where are we heading to? To have some clues about where this rapid (r)evolution will take us as computer scientists and engineers of complex systems, it might be worthwhile to take a look back to the evolution of human societies. As a matter of fact, recent theories interpret biological systems as information-based systems evolving in complexity through a handful of fundamental transitions [18]. On the other hand, artificial, human-built systems are roughly taking a similar path, growing in complexity through quantum leaps – Ubiquitous Computing possibly being one of them. It is then easy to think that some portions of the history might repeat: that some of the striking forces driving biological and social evolution might take new forms and drive the evolution of artificial systems as well; that some of the results of biological and social evolution might find their counterparts in the entities and structures of forthcoming human-built systems.

Latest results from both biological [18] and historical sciences [14] tell us that, after millions years of biological evolution, cultural and *social evolution* has become predominant – where social evolution basically means organisational changes in the social structure. Also, the role of the *environment* is reckoned as a dynamic and active force that drives the evolution of human societies, and also affects the ability of both individuals and social structures to survive, grow, and possibly overcome the others. According to [14], the environment where individuals and societies live determines which kinds of individuals survive and which kinds of societies / organisations prevail, typically by making resources more or less available, or by imposing constraints, limitations, and (vital) needs. While in the short / medium term, a more or less successful interaction of individuals and societies with the environment determines their success or failure, in the long term the environment is to be seen as one of the main factors (if not the main one) determining the successful evolution of individual and societies in the global setting.

On the other hand, the environment may evolve over time, and its evolution is partially under human's control. Depending on their understanding of the environment and of its dynamics, individuals and societies might plan and act so as to drive environment changes over time, and possibly survive and grow according to their ability in modeling and building / modifying their own environment.

So, while physical laws and phenomena (as environment emerging dynamics) are out of the human reach, the ability to model the environment and its dynamics, along with some goal-oriented activity and planning capability, enables humans and societies to induce some controlled change making the environment fit more their needs and desires, thus partially driving the environment evolution.

So, some preliminary questions already call for an answer: Who plays individuals and societies in the artificial systems designed and built by humans – today and tomorrow? What is environment, in the era of the Internet and Ubiquitous Computing?

12.1.2 A Look Forward to Agent Societies

To face the ever-growing complexity of artificial systems, increasingly powerful abstractions have been introduced in computer science (and related areas), which have gradually taken the software engineering (SE) field, and become best practice in the years: structured programming languages, modules, objects, components. Today, research on multiagent systems (MAS) is emerging as a fertile ground for powerful and expressive abstractions and tools, making it possible to face the new challenges in terms of system complexity. Agents are autonomous, goal-oriented, pro-active entities, organized in societies, and situated within an environment: notions like *agent*, *role*, *organization*, *society*, *environment* are the basic bricks of what is called Agent-Oriented Software Engineering (AOSE) [9, 38]. Even though MAS might seem now far from becoming SE mainstream, agent technology is already somehow ubiquitous, and the continuous progress in agent research and systems seems to encourage us to think that MAS are the next step in the evolution of artificial systems.

There seems to be no apparent need to advocate any anthropomorphic principle to reason about agents and their role in the future of artificial systems. Analogy is enough to drive comparisons and possibly help us foreseeing the evolution of MAS. More than ten thousands years have seen humans as individuals autonomously acting in an unpredictable and possibly hostile environment, trying to achieve their own goals by interacting with other individuals – cooperating, competing, coordinating – according to complex interdependency patterns. Humans have evolved as situated, speaking, specialisable, and context-aware entities, co-existing and interacting within societies, deeply immersed within their environment, which they represent, learn from, and try to affect / change for their own purposes. Also, according to [14], (successful) societies built by humans are open to change, feature some form of "social learning", and react to environment pressure and changes. Their organization fits a precise scale, and does not scale up: e.g., it can be observed that peer-to-peer organisations scale up to a maximum of 80-100 members, then fail. So, social organization is forced to dynamically evolve over time, to face the challenges posed by its own growth, by other competing human groups, and by the environment as well.

In agent arena, the above considerations still basically apply: the above sentences roughly holds for humans as well as for agents, and for human societies as well as for MAS. Analogy is not so strong, instead, when we take into account the notion of environment. In fact, it is still true that only a limited part of the environment falls within the reach of the control of individuals (agents/humans) and societies, and that in order to partially control the environment evolution over time the ability to model the environment and its dynamics is required, along with some goal-oriented activity and planning capabilities. Also, in agent systems, too, there exist non-controllable forces (e.g., the market, determining the survival of a MAS application) and maybe hostile, evil entities (e.g., virus, ill-driven organisations), that make agent environment a partially unpredictable factor in its very essence. On the other hand, however, we as humans are by definition the designers and builders of artificial systems – agent ones, in particular. So, the main difference to be pointed out here is our role as humans with respect to human and agent societies. While human beings play the role of individuals within human societies, with limited control on their environment, humans might act as *gods* with respect to agent societies. There, part of the environment is still out of the individual (agent) control, but is mostly defined and built according to computer scientists' and engineers' intentions and specifications: what is out of the control of the agent could be within the reach of the human.

189

The need for suitably powerful abstractions for modeling and engineering the agent environment as a first-class entity is then clear. What might not be so clear is that representing the environment and its dynamics in agent systems (and in artificial systems in general) is a twofold problem, depending on which viewpoint we take: either the agent, or the human one. So, the following fundamental questions at least require an answer: Which abstractions are the most expressive and manageable for computer scientists and engineers to model, build, and control the environment for agent systems? And, which abstractions are the best suited for agents (in particular intelligent ones) to enable them to perceive, understand, and possibly affect their environment according to their needs, desires and goals?

12.2 CONTEXTS EVERYWHERE

Literally speaking, a *context* is what comes along (both in a spatial and temporal sense) with a "text", that is, a collection of signs or symbols, and possibly conveys further elements to interpret and give meaning to the text itself. In general, the notion of *context* is relevant and used within several research areas such as language, philosophy, logic, and artificial intelligence. In the field of natural language, *context-dependent interpretation* is a notion of outmost importance, encompassing the dependency of interpretation on both the linguistic and the non-linguistic environment where words and sentences occur. In the same field, *computational models of context* have been developed, for instance for natural language generation. Furthermore, several different notion of context have been defined over time in logic, computer science, and artificial intelligence. The collection [3] reports on contextual deontic logic, contextual learning, contextual interpretation of objects in a semantic network, and many other examples.

In the area of computer science, contexts have often be studied as a means to encapsulate knowledge (whether factual or procedural) or beliefs, and to combine them incrementally and dynamically to model many different notions related to knowledge, beliefs, and communication. For instance, in the programming language field, *contextual logic programming* have promoted the notion of context to a first-class entity, as the dynamically evolving set of the axioms determining the truth value of logic formulae [12].

Most recently, the issue of *context-aware computing* has become mainstream in the human-computer interaction (HCI) area, where it has been recognized that (i) there is a need for models, methods and tools to capture the setting where interaction between humans and computers occurs, (ii) the notion of context is the most obvious candidate to provide the required conceptual foundation, but (iii) the notion of context is still ill-defined [13] – or, at least, no consensus can be reached about what a context precisely is [37].

In general, contexts are typically used to model the effect of the environment – in its most general acceptation, including the spatial and temporal interpretations of the term – on the interaction and communication occurring amongst active (and typically intelligent) entities, such as humans or artificial agents. Also, contexts typically account for *situatedness* of the interaction and communication, coming to say that interaction and communication cannot be understood outside of the environment where they occur – out of their context, in other words. So, in spite of the many different acceptation of the term within the many heterogeneous research areas where it is commonly used, the notion of context is the most suitable candidate to be used as the first-class abstraction for modeling and engineering the environment in agent systems.

12.3 OBJECTS VS. AGENTS: THE ISSUE OF CONTROL

Since the notion of context occurs in so many different areas, and in computer science as well, one may argue that in a field like object-oriented languages and systems, for instance, the need for explicitly modeling the environment seemingly did not emerge so far, at least not so clearly. So, what are "object societies", and why are they so different from agent societies, that a notion of "object context" has not been developed, yet?

Among the many possible answers, the most convincing one relates to the issue of control. In object-oriented (OO) systems, control flows along with data: as the reification of message passing, method invocation couples control and data, providing a single way to drive control along an OO system. Roughly speaking, design is *control-oriented* in OO systems, where control is outside objects, and the (human) designer works as a sort of "central control authority" – the *control god*, in other words.

If this simplifies system design at the small scale, it nevertheless makes design a non-trivial effort at the large scale, requiring engineering discipline to be effectively managed. The notion of object *interface* provides the form of discipline required, by constraining inter-object interaction, in terms of both control and data passing. Though the interface, control is passed to an object in a controlled way, and crosses object boundaries from outside in – when a method of the interface is invoked –, then from inside out – when the invoked method returns control to the caller. By describing the observable behaviour of objects, interfaces give discipline to object interaction and altogether provide the designer with a complete description of the interaction environment at design time. So, "social interaction" among objects is then limited and disciplined by interfaces, and governed by the human designer. In some sense, the environment surrounding an object in an application consists of all the other application objects, as described by their interfaces – the collection of the object interfaces could represent the context where the object lives and interacts. As a result, roughly speaking, while interfaces are enough for human designers to govern OO systems, objects are not powerful enough to require the development of explicit notions for object society and environment.

Instead, agent systems are not control-oriented, but *goal-oriented*. Agent systems have no central control authority (no control god), instead each agent is an independent *locus* of control: control is inside agents, and the agent goal drives the control – each agent works as its own control god. So, since data do not necessarily flow with control, control and data are uncoupled in agent systems. As any non-trivial form of decoupling, this makes system engineering simpler. In fact, the human designer is to some extent free of the burden of explicitly managing control in agent systems, and can in principle focus on designing agent system as information-oriented at the highest level of abstraction.

However, given that AOSE promotes agents as useful abstractions for modeling and engineering large complex systems [28, 27], the need for a disciplined environment for agent systems emerges clearly in the same way as in the case of OO systems.

191

This is even more true for applications in the Ubiquitous Computing scenario [36]. There, in fact, the huge number, diversity, and dynamics of the environments where agents would be called to (inter)act could give raise the same sort of complexity that already emerged in the robotics research – like the *frame problem*, or the *qualification problem*. That is, roughly speaking: Which actions is the agent actually allowed to perform at any place and time? Which effect on the surrounding environment will agent actions really have? Which knowledge about the environment is relevant for an agent to effectively plan and act? How and to what extent can an agent be ensured that its knowledge about the environment is at any time consistent with its actual state?

As a result, when we try to devise out an agent notion that could roughly correspond to object interface, we are led back to some agent-related notion of *context*, which could work as a limited yet effective representation of the agent environment, and also allow agent interaction to be properly disciplined. Also, the required engineering discipline essentially concerns agent interaction – within societies, and with the environment. So, since by definition *coordination* is managing [17] and constraining [35] interaction, it seems almost clear that the notion of context we need in complex agent systems is essentially that of an *agent coordination context*.

Even though it would play a role corresponding to object interfaces in object systems in some sense (that is, according to the line of reasoning developed in this section), an agent coordination context is obviously not an "agent interface" – no way. In its most general acceptation, an agent cannot be invoked: control is within agents, and does not cross the agent boundaries. An agent could indeed be designed as a service provider, but it is autonomous by definition: in principle, it would provide services on its own deliberation, and could then at any time refuse to do so – differently than objects, "agents can say no" [21].

However, object systems of today could tell us more about what an agent coordination context could be – more precisely, component-based systems could. In fact, component-based systems typically come along with a notion of *infrastructure*, providing *services* to application objects, like transparent access to distributed resources, directory services, name services. These infrastructures (like CORBA [22], DCOM [33], Jini [34]) embody some implicit definition of the environment for components and applications, by defining how components could enter a new environment, how could they identify themselves and the other participants, how could they locate, access or make available resources, and so on. Notions like interface description language and inspectable interface suggest us that an agent coordination context should be conceived not only as a tool for human designers, but also as a run-time abstraction provided as a service to agents by a suitable infrastructure.

Finally, the ever-growing capabilities of the abstractions in artificial systems is changing the roles they play. Expert systems are used in the diagnosis of errors and critical situations, objects are used to automatically balance loads in parallel systems, agents are used for network management, just to quote some. So, while we take as granted the meta-level roles that humans play in un artificial system (as designers, developers, deployers, maintainers, actors, final users), we often do not take into account as well that some of these roles – and perhaps in some future all of them – could be played by the components of an artificial system. This is for instance one of the concerns of the Semantic Web project [2], which basically considers software agents as alternative (to humans) users of Web pages. As a consequence, features of abstractions and tools for the engineering of artificial systems require more and more to be accessible to both humans and artificial system components: so, for instance, if we require that an agent coordination context be inspectable, this should mean that can be inspected by a human (say, the system manager) and by an intelligent agent as well.

12.4 AGENT COORDINATION CONTEXT

Generalizing upon the recently introduced notion of context-dependent coordination [6], this paper suggests the notion of *agent coordination context* as a means to model and shape environment in agent systems. As discussed in the previous section, this notion is expected to play in some sense the same role within agent systems as the notion of object interface in object systems: that is, providing a discipline for interaction, and some pattern for managing control. Since, roughly speaking, control is outside objects, but inside agents, an agent coordination context is indeed meant to provide a sort of "boundary description" as an object interface does – but from inside out (an agent), rather than from outside in (an object). As an interface, however, a coordination context provides decoupling between agents and their environment, so that the agent can in principle be designed and developed independently of other agents, resources, and services populating the MAS environment.

More precisely, a coordination context should

- work as a model for the agent environment, by describing the environment where an agent can interact, and

- enable and rule the interactions between the agent and the environment, by defining the space of the admissible agent interactions

As a tool to model the environment, an agent coordination context can serve a twofold purpose, according to the viewpoint we take. From the agent viewpoint (more generally called the *subjective* viewpoint [32]), an agent coordination context should provide agents with a suitable representation of the environment where they live, interact, and communicate. Such a representation should obviously include some notion of *locality*, in both space and time – since a global and synchronous environment is not realistic in today application scenarios –, implicitly defining the where and when of agent interactions. Obviously, a coordination context should contain all the information about the entities (agents, services, resources) an agent could interact with, along with a description of the admissible way of interaction. This obviously assumes that a *context representation language* is defined and used: it could be a first-order logic language, an XML-based ones, or whatever – until it is expressive enough to fully describe an agent environment and all the admissible agent interactions.

From the human designer viewpoint (more generally called the *objective* viewpoint [32]), an agent coordination context should provide a framework to express the interaction within a MAS as a whole. More precisely, coordination contexts define the space of MAS interaction, that is, the admissible interactions occurring among the agents of a MAS, and between the agents of a MAS and the MAS environment. In particular, as far as inter-agent communication is concerned, coordination contexts model how the environment affects interpretation of agent communication acts.

As a tool to enable and rule agent interaction with the environment, an agent coordination context can again serve a twofold purpose. From the agent viewpoint, the coordination context enables in principle agents to perceive the space where they act and interact, reason about the effect of their actions and communications, and possibly affect the environment to accomplish their own goals. From the viewpoint of a human engineer, coordination contexts would allow engineers to encapsulate rules for governing applications built as agent systems, mediate the interactions amongst agents and the environment, and possibly affect them so as to change global application behaviour incrementally and dynamically. Coordination contexts allow then a form of *prescriptive coordination* to be enforced, constraining from the design stage to the run time the space of agent interaction. By the way, it should be observed that only the capability to ensure that harmful behaviors could be prevented, and that only some admissible courses of action could be taken by agents, would make a host infrastructure provide open spaces where agents can act according to their own deliberation and will, free to accomplish their own goals, without constraining their model or behavior *a priori*. From this viewpoint, then, prescriptive coordination seems to be a sort pre-condition to agent autonomy, rather than an obstacle to it.

As observed in the previous section, this also calls for a suitable run-time support, embodying coordination contexts as run-time abstractions provided by the agent infrastructure. Even more, by their very nature, agent coordination contexts could be used to mediate between the needs of a MAS application and its host, by actually modeling agent infrastructures (their structure and organization, the resources and services they make available to agents) in terms of coordination contexts. When suitably supported as run-time abstractions, coordination contexts should be dynamically *configurable* and *inspectable* by both agents and humans. Configurability would allow a MAS to evolve at run time, by suitably adapting its behavior to changes – expectedly a major requirement in complex and highly dynamic scenarios such as the Ubiquitous Computing one [36]. Inspectability would allow both humans and intelligent agents to reason about the current laws of coordination as represented and embodied within coordination contexts, and to possibly change them by properly re-configure coordination contexts according to new application needs. Finally, since coordination models have been recognized as a basis to enforce social rules [8], and that contexts are easily interpreted as social constructs [1], coordination contexts have the potential to be exploited for the engineering of social order within agent societies [7].

12.5 EXAMPLES

Given the seminal nature of this notes, it is obvious that the notion of agent coordination context has lots of elements and details to be defined and made more precise before it becomes actually implementable and usable. As a result, cases of agent coordination contexts can be given only in terms of either a conceptual example, or a model of coordination where aspects of a coordination context can be suitably devised out at the proper level of abstraction.

12.5.1 The Control Room Metaphor

The *control room metaphor*, as defined in the following, provide a conceptual example of an agent coordination context. According to this metaphor, an agent entering a new environment is assigned its own control room, which is the only way in which it can perceive the environment, as well as the only way in which it can interact. There, *admissible agent inputs* are represented as *lights* (for time-discrete inputs) and *screens* (for time-continuous inputs), while *admissible agent output* are represented as *buttons* (for time-discrete outputs) and *cameras* (for time-continuous outputs). Thus, for instance, bi-directional continuous communication with another agent could be carried on through a dedicated pair screen-camera.

How many input and output "devices" are available to an agent, of what sort, and for how much time – this is what defines the control room *configuration*, that is, the specific agent coordination context. Such a configuration, representing a sort of environment interface for the agent, is at any time prescriptive, in that it fully describes all the admissible interactions for an agent: however, it might change over time whenever needed. In fact, the initial control room configuration is subject to preliminary negotiation between the agent and the hosting node. Then, it could be dynamically modified according to changes in the needs of either the agent or the hosting node providing the coordination context. For instance, due to sudden lack in resources, a control room could be reduced by the hosting infrastructure to the empty context (no interaction allowed, in other words), implicitly forcing the agent to move elsewhere in order to be able to perform some meaningful action. Also, one may think to associate a cost (per time, per use) to every devices made available by a configuration, so that negotiations for the initial control room configuration, or for a subsequent change in the configuration, would also take the economic issue into account.

12.5.2 Coordination Contexts in the TuCSoN Coordination Model

A coordination model as defined in [31] provides a framework and a technology to shape and govern the space of agent interaction. A meaningful example of a coordination model and technology is TuCSoN [29]: there, the coordination space is defined as a collection of local interaction spaces, and agent interaction is mediated through programmable tuple spaces called *tuple centers* [25]. Each TuCSoN hosting node is equipped with an infrastructure providing agents with a set of tuple centers as run-time coordination abstractions to be used to interact with other agents and with the local environment as well. Tuple centers are Linda-like tuple spaces [16] enhanced with the notion of behavior specification, expressed in terms of the first-order logic specification language ReSpecT [24]. Each tuple center is independently programmed through its own ReSpecT specification, defining its behavior in response to communication events. As a result, a tuple center encapsulates the laws of coordination in terms of a ReSpecT specification, which are both inspectable and dynamically configurable.

In terms of a coordination context, TuCSoN provides each agent with a view of the space of interaction that includes a collection of coordination media, namely the tuple centers – but is not necessarily limited to that. On the one hand, in fact, the TuCSoN infrastructure is not prescriptive at all: any interaction pattern other than the tuple center-mediated one is in principle available to agents. So, for instance, two agents could choose to interact directly and bypass tuple center mediation. On the other hand, since tuple centers are provided as *coordination services* by the TuCSoN infrastructure, the deliberate choice of the agents to exploit them is enough to bound and constrain their interaction: until agents interact through tuple centers, they will follow the coordination laws that tuple centers embody. Also, the coordination context provided by TuCSoN is in general inspectable and dynamically configurable by both humans and (intelligent) agents: humans can access and modify tuple centers through inspectors, made available by the infrastructure as development and deployment tools, whereas agents are provided with a set of special meta-level primitives to inspect and change tuple center behavior specifications. Since tuple centers encapsulate both the state of the communication and the laws of coordination as first-order logic clauses, they can be in principle exploited by both humans and agents to monitor a MAS evolution over time, and to possibly change MAS behavior dynamically by suitably affecting their ReSpecT behaviour specification.

12.6 RELATED APPROACHES & OPEN ISSUES

The notion of *context-dependent coordination* was first introduced in [6]. There, as in TuCSoN [29], a multiplicity of different programmable tuple spaces (tuple centers [25]) are used to provide mobile agents with different coordination contexts, independently programmable, and configurable by agents according to their own application needs. Here too, however, prescriptiveness of coordination is bound to either the choice of agents to interact through tuple centers only, or the ability of the infrastructure to super-impose tuple center-mediated interaction to agents – however, this issue is not explicitly discussed in the paper.

A better defined notion of coordination context is instead introduced by *law-governed interaction* [19]. There, a set of policy-independent trusted controllers enforces coordination laws as access policies to tuple spaces locally to agents. Co-ordination laws are then embodied locally to controllers, which basically work as a sort of coordination contexts.

Even though they do not discuss explicitly a notion of context, agent-oriented methodologies are more and more stressing the role of the environment in the engineering of complex systems. Methodologies like Gaia [39] and SODA [23] emphasize the idea that both agent societies and environment have to be handled as first-class entities. In particular, SODA is explicitly meant to exploit coordination models and technologies for the engineering of the MAS aspects related to agent societies and environment.

Given the seminal nature of these notes, several issues remain open to further research. The first concerns the problem of situatedness and explicit representation of context [40]. While the notion of context representation language naturally suggests a symbolic approach to environment representation, it is not yet clear how much the very notion of situatedness would cope well with symbolic representation of the environment [4]. Also, it is not clear whether communication should be handled and represented as a special kind of action, or as an independent modality of interaction [15].

The notion of prescriptive coordination discussed in previous sections raise the issue of security, in particular in the context of open systems. Given that the main goal of security is, roughly speaking, to manage agent/environment interactions so as to prevent possibly harmful behaviors, security can in principle be intepreted as a coordination issue [10, 5]. So, how can an agent coordination context represent and embody the level of security required by today systems? And to what extent can security policies be defined and enforced, without harming the fundamental notion of agent autonomy? Some work in this direction already began, even though it is still quite preliminary [11].

Last but not least, coordination contexts could provide systems with an economy-oriented view of the environment. Information has a cost, interaction and communication have a cost, too: access to resources and services, as well as to communication media, is not to be taken as free to agents and applications in general. Also, coordination adds a value per se, which should be properly accounted for. So, a coordination context could in principle be used also to provide agents with an economy model of the environment, even though it is not clear how costs should be represented in general. For instance, an agent entering a new environment could initially contract to buy a cheap and "weak" coordination context (with only limited access to resources), and only after a preliminary exploration phase deliberate whether to leave the place (having found it not interesting at all) or to stay there, possibly extending its local interaction capabilities by negotiating (and paying for) a more powerful coordination context.

12.7 CONCLUSIONS

The study of the evolution of human societies tells us that we cannot even understand the evolution of individuals and groups in any non-trivial setting without explicitly taking the environment into account. The notion of context is already extensively used in several research fields to model the effect of the environment upon individuals and organizations. Also, a notion corresponding to object interface is required in agent systems to discipline and engineer the space of agent interaction – both agent-to-agent and agent-to-environment interactions. Then, the notion of *agent coordination context* was introduced here as both a representation tool and a run-time abstraction meant to handle agent environment as a first-class entity in the modeling and engineering of complex agent systems.

It is anyhow clear that any feasible and manageable notion of agent coordination context requires ontologies, abstractions, and languages to represent agent interaction within an environment, as well as run-time technologies, tools, and methodologies to handle the environment as a first-class entity in the modeling and engineering of agent systems. Most of these issues are likely to be addressed in the research work that will follow these seminal notes.

12.8 ACKNOWLEDGEMENTS

First, I would like to thank MIUR (the Italian "Ministero dell'Istruzione, dell'Università e della Ricerca") and Nokia Research Center, Burlington, MA, USA, for partially supporting this work. Also, let me thank the organizers of the Workshop on Internet Process Coordination and Ubiquitous Computing, and all the attendants as well, that stimulated the development of the concepts and ideas presented here.

Finally, I am deeply grateful to Alessandro Ricci, whose continuous flow of materials, ideas, remarks and criticisms was of invaluable help in improving the various version of this article.

REFERENCES

1. Varol Akman. Rethinking context as a social construct. *Journal of Pragmatics*, 32(6):743–759, 2000.

2. Tim Berners-Lee, James Hendler, and Ora Lassila. The semantic web. *Scientific American*, May 2001.

3. Pierre Bonzon, Marcos Cavalcanti, and Rolf Nossum, editors. *Formal Aspects of Contexts*, volume 20 of *Applied Logic Series*. Kluwer Academic Publishers, 2000.

4. Rodney A. Brooks. *Cambrian Intelligence: The Early History of the New AI*. MIT Press, 1999.

5. Ciarán Bryce and Marco Cremonini. Coordination and security on the Internet. In Omicini et al. [30], chapter 11, pages 274–298.

6. Giacomo Cabri, Letizia Leonardi, and Franco Zambonelli. Context-dependency in Internet-agent coordination. In Omicini et al. [28], pages 51–63.

7. Cristiano Castelfranchi. Engineering social order. In Omicini et al. [28], pages 1–18.

8. Paolo Ciancarini, Andrea Omicini, and Franco Zambonelli. Multiagent system engineering: The coordination viewpoint. In Nicholas R. Jennings and Yves Lespérance, editors, *Intelligent Agents VI. Agent Theories, Architectures, and Languages*, volume 1757 of *LNAI*, pages 250–259. Springer-Verlag, 2000. 6th International Workshop, ATAL'99, Orlando, FL, USA, 15–17 July 1999, Proceedings.

9. Paolo Ciancarini and Michael J. Wooldridge, editors. *Agent-Oriented Software Engineering*, volume 1957 of *LNCS*. Springer-Verlag, 2001. 1st International Workshop, AOSE 2000, Limerick, Ireland, 10 June 2000, Revised Papers.

10. Marco Cremonini, Andrea Omicini, and Franco Zambonelli. Multi-agent systems on the Internet: Extending the scope of coordination towards security and topology. In Francisco J. Garijo and Magnus Boman, editors, *Multi-Agent Systems Engineering*, volume 1647 of *LNAI*, pages 77–88. Springer-Verlag, 1999. 9th European Workshop on Modelling Autonomous Agents in a Multi-Agent World, MAAMAW'99, Valencia, Spain, 30 June – 2 July 1999, Proceedings.

11. Marco Cremonini, Andrea Omicini, and Franco Zambonelli. Coordination and access control in open distributed agent systems: The TuCSoN approach. In António Porto and Gruia-Catalin Roman, editors, *Coordination Languages and Models*, volume 1906 of *LNCS*, pages 99–114. Springer-Verlag, 2000. 4th International Conference, COORDINATION 2000, 11–13 September 2000, Limassol, Cyprus, Proceedings.

12. Enrico Denti, Antonio Natali, Andrea Omicini, and Francesco Zanichelli. Robot control systems as contextual logic programs. In Christoph Beierle and Lutz Plümer, editors, *Logic Programming: Formal Methods and Practical Applications*, volume 11 of *SCSAI*, pages 343–379. Elsevier Science, 1995.

13. Anind K. Dey, Daniel Salber, and Gregory D. Abowd. A conceptual framework and a toolkit for supporting the rapid prototyping of context-aware applications. In *Human-Computer Interaction* [20], pages 97–166.

14. Jared Diamond. *Guns, Germs, and Steel: The Fates of Human Societies*. W.W. Norton & Company, 1st edition, March 1997.

15. Peter Gärdenfors. The pragmatic role of modality in natural language. In Paul Weingartner, Gerhard Schurz, and Georg Dorn, editors, *The Role of Pragmatics in Contemporary Philosophy*, pages 78–91. Holder-Pichler-Tempsky, Vienna, 1998. 20th International Wittgenstein Symposium, Proceedings.

16. David Gelernter. Generative communication in Linda. *ACM Transactions on Programming Languages and Systems*, 7(1):80–112, January 1985.

17. Thomas Malone and Kevin Crowstone. The interdisciplinary study of coordination. *ACM Computing Surveys*, 26(1):87–119, 1994.

18. John Maynard Smith and Eörs Szathmáry. *The Origins of Life: From the Birth of Life to the Origins of Language*. Oxford University Press, 1st edition, May 1999.

19. Naftaly H. Minsky, Yaron M. Minsky, and Victoria Ungureanu. Safe tuplespace-based coordination in multiagent systems. In *Applied Artificial Intelligence* [26], pages 11–34.

20. Thomas P. Moran and Paul Dourish. Context-aware computing. *Human-Computer Interaction*, 16(2-4), 2001. Special Issue.

21. James J. Odell, H. Van Dyke Parunak, and Bernhard Bauer. Representing agent interaction protocols in UML. In Ciancarini and Wooldridge [9], pages 121–140.

22. OMG. CORBA home page. http://www.omg.org/corba/.

23. Andrea Omicini. SODA: Societies and infrastructures in the analysis and design of agent-based systems. In Ciancarini and Wooldridge [9], pages 185–193.

24. Andrea Omicini and Enrico Denti. Formal ReSpecT. In Agostino Dovier, Maria Chiara Meo, and Andrea Omicini, editors, *Declarative Programming – Selected Papers from AGP'00*, volume 48 of *Electronic Notes in Theoretical Computer Science*, pages 179–196. Elsevier Science B. V., 2001.

25. Andrea Omicini and Enrico Denti. From tuple spaces to tuple centres. *Science of Computer Programming*, 41(3):277–294, 2001.

26. Andrea Omicini and George A. Papadopoulos. Coordination models and languages in AI. *Applied Artificial Intelligence*, 15(1):1–103, 2001. Special Issue.

27. Andrea Omicini, Paolo Petta, and Robert Tolksdorf, editors. *Engineering Societies in the Agents World II*, volume 2203 of *LNAI*. Springer-Verlag, 2001. 2nd International Workshop, ESAW'01, Prague, Czech Republic, 7 July 2001, Revised Papers.

28. Andrea Omicini, Robert Tolksdorf, and Franco Zambonelli, editors. *Engineering Societies in the Agents World*, volume 1972 of *LNAI*. Springer-Verlag, 2000. 1st International Workshop, ESAW'00, Berlin, Germany, 21 August 2000, Revised Papers.

29. Andrea Omicini and Franco Zambonelli. Coordination for Internet application development. *Autonomous Agents and Multi-Agent Systems*, 2(3):251–269, 1999. Special Issue: Coordination Mechanisms for Web Agents.

30. Andrea Omicini, Franco Zambonelli, Matthias Klusch, and Robert Tolksdorf, editors. *Coordination of Internet Agents: Models, Technologies, and Applications*. Springer-Verlag, March 2001.

31. George A. Papadopoulos. Models and technologies for the coordination of Internet agents: A survey. In Omicini et al. [30], chapter 2, pages 25–56.

32. Michael Schumacher. *Objective Coordination in Multi-Agent System Engineering – Design and Implementation*, volume 2039 of *LNAI*. Springer-Verlag, 2001.

33. Roger Sessions. *COM and DCOM: Microsoft's Vision for Distributed Objects*. John Wiley & Sons, December 1997.

34. Jim Waldo. The Jini architecture for network-centric computing. *Communications of the ACM*, 42(7):76–82, 1999.

35. Peter Wegner. Why interaction is more powerful than computing. *Communications of the ACM*, 40(5):80–91, 1997.

36. Mark Weiser. Hot topic: Ubiquitous computing. *IEEE Computer*, 26(10):71–72, October 1993.

37. Terry Winograd. Architectures for context. In *Human-Computer Interaction* [20].

38. Michael J. Wooldridge, Paolo Ciancarini, and Gerhard Weiss, editors. *Agent-Oriented Software Engineering II*, volume 2202 of *LNCS*. Springer-Verlag, 2002. 2nd International Workshop, AOSE 2001, Montreal, Canada, 29 May 2001, Revised Papers & Invited Contributions.

39. Franco Zambonelli, Nicholas R. Jennings, Andrea Omicini, and Michael J. Wooldridge. Agent-oriented software engineering for Internet applications. In Omicini et al. [30], chapter 13, pages 326–346.

40. Tom Ziemke. Embodiment and context. In *European Conference on Cognitive Science, ECCS'97, Workshop on Context*, Manchester, UK, 9–11 April 1997.

Author(s) affiliation:

- **Andrea Omicini**

  DEIS, Università di Bologna, Sede di Cesena
  via Rasi e Spinelli 176, 47023 Cesena, FC, Italy
  Email: aomicini@deis.unibo.it

13

Thin Middleware for Ubiquitous Computing

Koushik Sen
Gul Agha

13.1 INTRODUCTION

As Donald Norman put it in his popular book *Invisible Computers* [12], "a good technology is a disappearing technology." A good technology seamlessly permeates into our lives in such a way that we use it without noticing its presence. It is invisible until it is not available. Ever since the invention of microprocessors, many computer researchers have strived to make computer technology a "good" technology.

Advances in chip fabrication technology have reached a point where we can physically make computing devices disappear. Bulkier machines have given way to smaller yet more powerful personal computers. It has become possible to implant a complete package of a microprocessor with wireless communication, storage, and a sensor on a cubic millimeter silicon die [8]. Specialized printers print out computer chips on a piece of plastic paper [6]. Computer chips are woven into on a piece of fabric [9]. "Smart labels" (a.k.a passive RFID tags) [7] will soon be attached to every product in the market.

The growth of devices with embedded computers will provide task-oriented, simple services which are highly optimized for their operating environment. More user oriented, human friendly services may be created by networking the embedded nodes, and coordinating their software services.

Composing existing component services to create higher-level services has been promoted by CORBA, DCOM, Jini, and similar middleware platforms. However, these middleware services were designed without paying much attention to resource management issues pertinent to embedded nodes: the middlewares tend to have large footprints that do not fit into the typically small memories of tiny embedded computers. Thus there is a need for a middleware which allows components to be glued together without a large overhead.

Embedded nodes are autonomous and self contained. They have their own state and a single thread of control, and a well defined interface for interaction. Interaction between nodes is asynchronous in nature. The operating environment for these devices may be unreliable. For these reasons, the interaction between different devices must be arms-length – the failure of one device should not affect another. For example, consider an intelligent home where the clock is networked to the coffee maker and an alarm also triggers the coffee maker. An incorrectly operating coffee maker should not cause the alarm clock to fail to operate. The autonomy and asynchrony in the model we describe helps ensure such fault containment.

Embedded nodes are typically resource constrained – they have small communication range, far less storage, limited power supply, etc. These resource limitations have a number of consequences. Small memory means that not every piece of code that may be required over the lifetime of a node can be pre-loaded onto the node. Limited power supply means that certain abstractions, such as those requiring busy waiting to implement, may be too expensive to be practical.

We propose *thin middleware* as a model for network-centric, resource-constrained embedded systems. There are two aspects to the thin middleware model. First, we represent component services as *Actors* [1, 3]. The need for autonomy and asynchrony in resource-constrained networks of embedded nodes makes Actors an appropriate model to abstractly represent services provided by such systems. Service interactions in the thin middleware are modeled as asynchronous communications between actors. Second, we introduce the notion of meta-actors for service composition and customization. Meta-actors represent system level behavior and interact with actors using an event-based signal/notify model.

13.2 ACTORS

Actors [1, 3] were developed as a basis for modeling distributed systems. An actor encapsulates a state, a set of procedures which manipulate the state, and a thread of control. Each actor has a unique *mail address* and a *mail buffer* to receive messages. Actors compute by serially processing messages queued in their mail buffers. An actor waits if its mail buffer is empty. Actors interact by sending messages to each other.

In response to a message, an actor carries out a local computation (which may be represented by any computer program) and three basic kinds of actions (see Figure 13.1):

Send messages: an actor may *send* messages to other actors. Communication is point-to-point and is assumed to be weakly fair: executing a *send* eventually causes the message to be buffered in the mail queue of the recipient. Moreover, messages are by default asynchronous and may arrive in an order different from the one in which they were sent.

Create actors: An actor may *create* new actors with specified behaviors. Initially, only the creating actor knows the name of the new actor. However, actor names are first class entities which may be communicated in messages; this means

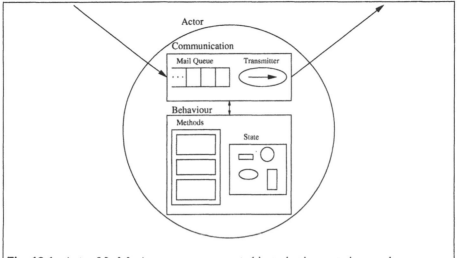

Fig. 13.1 **Actor Model:** Actors are concurrent objects that interact via asynchronous messages.

that coordination patterns between actors may be dynamic and the system is extensible.

Become ready to accept a message: The actor becomes *ready* to process the next message in its mail queue. If there is no message in its mail queue, the actor waits until a new message arrives and processes it.

Asynchronous message passing is the distributed analog of method invocation in sequential object-oriented languages. The *send* and *create* operations can be thought of as explicit requests, while the *ready* operation is implicit at the end of a method. That is, actors do not explicitly indicate that they are ready to receive the next message. Rather, the system automatically invokes *ready* when an actor method completes.

Actor computations are abstractly represented using *actor event diagrams* as illustrated in Figure 13.2. Two kinds of objects are represented in such diagrams: actors and messages. An actor is identified with a vertical line which represents the life-line of the actor. The darker parts on the line represent the processing of a message by the actor. The actor may create new actors (dotted lines) and may send messages (solid lines) to other actors. The messages arrive at their target actors after arbitrary but finite delay and get enqueued at the target actor's mail queue.

Note that the nondeterminism in actor systems results from possible shuffles of the order in which messages are processed. There are two causes of this nondeterminism. First, the time taken by a message to reach the target actor depends on factors such as the route taken by the message, network traffic load, and the fault-tolerance protocols used. Second, the order in which messages are sent may itself be affected by the processing speed at a node and the scheduling of actors on a given node. Nondeterminism in the order of processing messages abstracts over possible communication and scheduling delays.

203

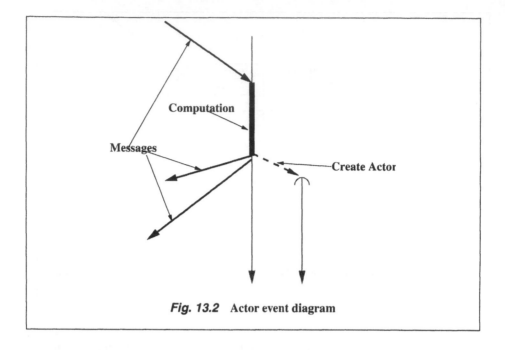

Fig. 13.2 Actor event diagram

The nondeterministic model of concurrency provides a loose specification. Properties expressed in this model state what *may*, or what *must*, eventually happen. In reality, the probability that, for example, a message sent at a given time will be received after a million years is practically infinitesimal. One way to express constraints on the arbitrary interleavings resulting from a purely nondeterministic model is by using a probabilistic model. In such a model, we associate a probability with each transition which may depend on the current state of the system. Our specifications then say something about when something may happen with a given probability. We discuss a probabilistic model in some more detail below.

13.2.1 Probabilistic Discrete Real-time Model

Traditional models of concurrent computation do not assume a unique global clock – rather each actor is asynchronous (for example, see [3, 2]). However, when modeling interaction of the physical world with distributed computation, it is essential to consider guarantees in real-time. Such guarantees are expressed in terms of a unique global time or wall clock and the behavior of all devices and nodes is modeled in terms of this reference time (for example, see [11, 14, 13, 10]). This amounts to a synchronous model of actors and it implies a 'tight coupling' in the implementation of actors; network and scheduling delays, as well as clock drift on the nodes, must be severely restricted.

In network embedded systems, a number of factors make a tight coupling in the implementation of actors infeasible. For example, the operation of some embedded devices may be unreliable, and message delivery may have nondeterministic delays due to transmission failures, collisions, and message loss. So in large network embed-

204

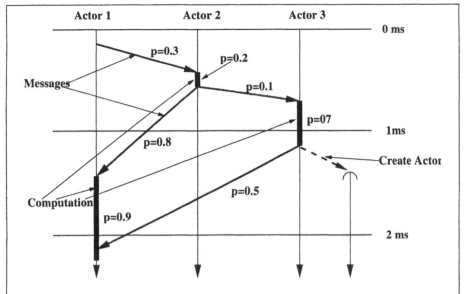

Fig. 13.3 **Probabilities** to capture nondeterminism in computation and communication

ded systems, it is not feasible to maintain a unique reference clock. The introduction of probability in the operations can be thought of as an intermediate synchronization model. In a probabilistic model, we assume that the embedded nodes agree on a global clock, but their drift from the clock is only probabilistically bound. Such probabilities replace the qualitative nondeterminism in computation and communication.

We assume the actors and the messages in transit form a *soup*. The components of the system follow a reference clock with some probability. The global time of the whole system (soup) advances in discrete time steps. The time steps can be compressed and stretched, depending on the kind of property we want to express. For example, the time step can be set to one second or it may be set to one millisecond. The global time of the system advances by one step when all the *actions* (computation and communication) that are possible in that time step have happened (see Figure 13.3). We associate a local clock with each actor and it advances with every global time step. However, it is reset to zero when the actor consumes a message. The clock remains zero when the actor is idle.

At a given time step an actor may be in one of three states:

- ready to process a message from its mail queue,

- busy computing, or

- waiting, because there is no message in mail queue.

If the actor is in either of the first two states, it can take the following actions:

- *complete* the computation in its current time step; or,

- *delay* its computation by one time step.

205

In the first case, the local clock of the actor remains same and so it is open to other actions in that time step. However, for the second action the local clock of the actor advances by one time step and hence, all possible actions of that actor get disabled for that time step. The two actions get enabled once the global time advances to the next time step. As a function of the state of the system, we associate different probabilities with each of the two actions.

Similarly, at a given time step a message can take three actions:

- it can get *enqueued* at the target actor,

- it can get *lost* and thus removed from the soup, or

- it can get *delayed*, in transit, by one time step.

If a message is delayed in transit, the local time of the message advances by one time step and so the message cannot take any more actions in that global time step. However, all the three actions will get enabled once the global time advances to the next step. Probabilities are associated with each action. The probabilities depend on factors such as the message density in the route taken by the message, the time for which it has been delayed (value of local clock of the message), and the number of messages sharing the same communication channel.

A computation path is defined as a sequence of states that the system has seen in the course of its computation. Note that the system retains the same computation paths as it would have in a nondeterministic model of concurrency. The probability that a particular finite sequence of states in a path will occur is obtained by multiplying the probabilities of all the actions in that sequence of states. Some of these probabilities will grow sufficiently small that they will no longer be relevant to the proof of some properties of our interest.

Using the above model, we can express properties of the form: "Within time t, the system will reach a state which satisfies a property π with probability p." For example:

- the alarm clock will ring at 7:00 a.m. with probability 0.99.

- the microwave will complete popping 95% of the popcorn by 10 a.m. with probability 0.98.

In implementing probabilistic timing specifications, one constrains the system level behavior which involves networks of heterogeneous nodes. A middleware provides a uniform interface to access such nodes. We represent the middleware itself as a collection of actors. The model we describe enables dynamic customizability of the execution environment of an actor in order to satisfy properties such as timing and security.

13.3 REFLECTIVE MIDDLEWARE

A key requirement for middleware is that it must enable dynamic customization – so that services can be pushed in and pulled out at runtime. This scheme of pushing-in and pulling-out of services allows the middleware to keep on a node only those services that are required by an application. The result is a light weight middleware.

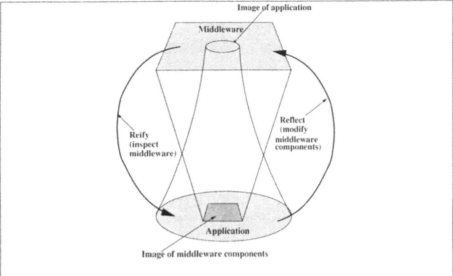

Fig. 13.4 **Reflection:** Application can inspect and modify middleware components.

Because an application may be aware of system level requirements for timing, security, or messaging protocols, it needs to have access to the underlying system. We support the ability of an application to modify its system level requirements by dynamically changing the middleware through the use of *computational reflection*.

A reflective middleware provides a representation of its different components to the applications running on top of it. The applications can *inspect* this representation and modify it. The modifications made to the components are immediately mirrored to the application. In this way, applications can dynamically customize the different components of the middleware through reflection (see Figure 13.4).

We use the *meta-actor* extension of actors to provide a mechanism of architectural customization [5]. A system is composed of two kinds of actors: base actors and meta-actors. Base actors carry out application-level computation, while meta-level actors are part of the runtime system (middleware) that manages system resources and controls the base-actor's runtime semantics.

13.3.1 Meta-architecture

From a systems point of view, actors do not directly interact with each other: instead, actors make *system method* calls which request the middleware to perform a particular action. A system method call which implements an actor operation is always 'blocking': the actor waits till the system signals that the operation is complete. Middleware components which handle system method calls are called *meta-actors*. A meta-actor executes a method invoked by another actor and returns on the completion of the execution. The requisite synchronization between an actor and its meta-actor is facilitated by treating the meta-actor as a passive object: it does not have its own thread of control. Instead, the calling object is suspended. In other words, an actor and its

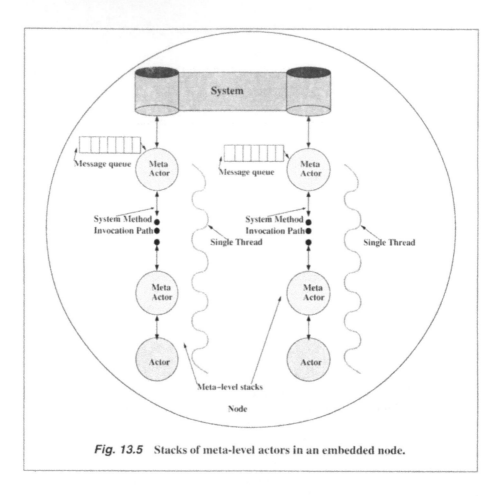

Fig. 13.5 Stacks of meta-level actors in an embedded node.

meta-actor are not concurrent – the latter represents the system level interpretation of the behavior of the former.

A meta-actor is capable of customizing the behavior of another actor by executing the method invoked by it. An actor customized in this fashion is referred to as the *base actor* relative to its meta-actor. To provide the most primitive model of customization a meta-actor can customize a single base-actor. However, multiple customizations may be applied to a single actor by building a *meta-level* stack, where a meta-level stack consists of a single actor and a stack of meta-actors (see Figure 13.5). Each meta-actor customizes the actor which is just below it in the stack. Messages received by an actor in a meta-level stack are always delegated to the top of the stack so that the meta-actor always controls the delivery of messages to its base-actor. Similarly messages sent by an actor pass through all the meta-actors in the stack.

We identify each operation of a base-actor as a system method call as follows.

- **send(*msg*):** This operation invokes the system method transmit with msg (msg is the message sent by the actor) as argument. If the actor has a meta-actor on its top it calls the transmit method of the meta-actor and wait for its

208

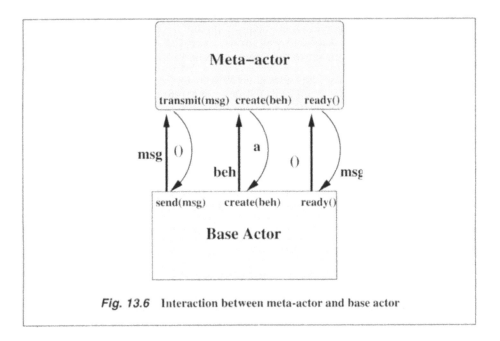

Fig. 13.6 Interaction between meta-actor and base actor

return. The method returns without any value. Otherwise, if the actor is not customized by a meta-actor, it passes the message to the system for sending.

- **create(*beh*):** This operation invokes the system method create with the given beh (beh is the behavior with which the newly created actor will be instantiated) as argument. If there is a meta-actor on top of the actor, it calls the create method of the meta-actor and waits for its return. The method returns the address a of the new actor. Otherwise, the actor passes the create request to the system.

- **ready():** The system method ready is invoked when an actor has completed processing the current message and is waiting for another message. If the actor has a meta-actor on top it calls the ready method of the meta-actor and waits for its return. The method returns a message to the base-actor. Otherwise, the actor picks up a message from its mail queue and processes it. Notice, there is a single mail-queue for a given meta-level stack.

The method call-return mechanism for different actor operations and the availability of a single queue for a meta-level stack makes the execution of a meta-level stack single threaded. So explicit scheduling of each actor in the stack is not required. The meta-actors behave as reactive passive objects which respond only when a system method is invoked by its base actor. The single thread implementation of a meta-level stack is important, as most of the embedded devices can have a single thread only. An example of such a embedded OS is TinyOS which runs on motes.

Every meta-actor has a default implementation of the three system methods. These implementations may be described as follows:

209

- **transmit(*msg*):** If there is a meta-actor on its top, it calls `transmit (msg)` method of that meta-actor and waits for it to return. Otherwise, it asks the system to send the message to the target and returns.

- **create(*beh*):** If the actor has a meta-actor at its top, it calls `create (beh)` method of that meta-actor and waits for the actor to return with an actor address. Otherwise, the actor passes the create request to the system and waits till it gets an actor address from the system. After receiving new actor address, the actor returns it to the base actor.

- **ready():** If there is a meta-actor on top of it, it calls `ready ()` method of that meta-actor and waits for it to return a message. Otherwise, the actor, by definition located at the top of the meta-level stack, dequeues a message from the mail queue. After getting the message, the actor returns the message to the base actor.

```
actor Encrypt(actor receiver) {        actor Decrypt() {

    // Encrypt outgoing                    // Decrypt incoming messages
    // messages if they                    // targeted for
    // are targeted to                     // base actor (if necessary)
    // the receiver                        method ready() {
    method transmit(Msg msg) {                Msg msg = ready();
        actor target = msg.dest;              if (encrypted(msg))
        if (target == receiver)                  return(decrypt(msg));
            target ← encrypt(msg);            else
        else                                     return(msg);
            target ← msg;                  }
        return;                        }
    }
}
```

Fig. 13.7 **Meta-Level Implementation of Encryption:** The `Encrypt` meta-actor intercepts `transmit` signals and encrypts outgoing messages. The `Decrypt` policy actor intercepts messages targeted for the receiver (via the `rcv` method) and, if necessary, decrypts an incoming message before delivering it.

As an example of how we may customize actors under this model, consider the encryption of messages between a pair of actors. Figure 13.7 gives pseudo-code for a pair of meta-actors which may be installed at each endpoint. The `Encrypt` meta-actor implements the `transmit` method which is called by the base-actor while sending a message. Within `transmit`, a message is encrypted before it is sent to its target. The `Decrypt` meta-actor implements the `ready` method which is called when the base actor is ready to process a message. Method `ready` decrypts the message before returning the message to the base-actor.

The abstraction of the middleware in terms of meta-actors gives the power of dynamic customization. Meta-actors can be installed or pulled out dynamically. This pushing in and pulling out of meta-actors by the application itself makes it capable of customizing the middleware. It also makes it possible to have only those middleware components which are required by services of the current application – facilitating our goal of thin middleware.

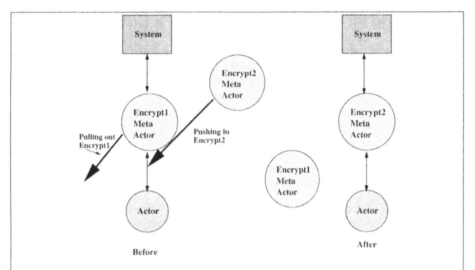

Fig. 13.8 Dynamic Customization: Pulling out and pushing in a new meta-actor for the implementation of encryption algorithm

13.4 DISCUSSION

We have described some preliminary work on a model of reflective middleware. We believe that further development of thin middleware will be central to the future integration of computing and the physical world [4]. However, many important problems have to be addressed before such an integration can be realized. We describe two areas to illustrate the problems. These areas relate, respectively, to the model and implementation of middleware.

A formal model of the interaction of the properties of actors and meta-actors has been developed in terms of a *two-level semantics* [15]. This model needs to be extended to its probabilistic real-time counterpart. For example, methods for composition of transition probabilities for actors and their meta-actors have not been developed.

More research is required in the implementation of thin middleware. Current implementation of reflective actor middleware has been based on high-level languages – which necessarily assume a large infrastructure. An alternate implementation would be in terms of a very efficient and small virtual machine which allows enforcement of timing properties. Related problems are incrementally compiling high-level code to such a virtual machine and supporting the mobility of actors executing on the virtual machine.

In our view, the solution to these and related problems define an ambitious research agenda for the coming decade.

211

13.5 ACKNOWLEDGMENTS

The research described here has been supported in part by the Defense Advanced Research Projects Agency (Contract numbers: F30602-00-2-0586 and F33615-01-C-1907). We would like to thank Nadeem Jamali and Nirman Kumar for reviewing previous versions of this paper and giving feedback.

REFERENCES

1. G. Agha. *Actors: A Model of Concurrent Computation*. MIT Press, 1986.

2. G. Agha. Modeling Concurrent Systems: Actors, Nets, and the Problem of Abstraction and Composition. In *17th International Conference on Application and Theory of Petri Nets*, Osaka, Japan, June 1996.

3. G. Agha, I. A. Mason., S. F. Smith, and C. L. Talcott. A foundation for actor computation. *Journal of Functional Programming*, 7:1–72, 1997.

4. Gul A. Agha. Adaptive middleware. *Communications of the ACM*, 45(6):30–32, June 2002.

5. M. Astley and G. Agha. Customization and composition of distributed objects: Middleware abstractions for policy management. In *Proceedings of the Sixth International Symposium of Foundations of Software Engineering*, pages 1–9, 1998.

6. S. B. Fuller, E. J. Wilhelm, and J. M. Jacobson. Ink-jet printed nanoparticle microelectromechanical systems. *Journal of Microelectromechanical Systems*, 11(1):54–60, 2002. http://www.media.mit.edu/molecular/projects.html.

7. AIM. Inc. http://www.aimglobal.org/technologies/rfid/.

8. J. M. Kahn, R. H. Katz, and K. S. J. Pister. Mobile Networking for Smart Dust. In *ACM/IEEE Intl. Conf. on Mobile Computing and Networking (MobiCom 99)*, Seattle, WA, August 1999. http://robotics.eecs.berkeley.edu/ pister/SmartDust/.

9. MIT Media Lab. http://lcs.www.media.mit.edu/projects/wearables/.

10. B. Nielsen and G. Agha. Semantics for an Actor-Based Real-Time Language. In *4th International Workshop on Parallel and Distributed Real-Time Systems (WPDRTS). Submitted*. Naval Surface Warfare Center Dahlgren Division/IEEE, April 1995. In conjunction with 10th IEEE Int. Parallel Processing Symposium (IPPS), Honolulu, Hawaii, USA.

11. B. Nielsen and G. Agha. Towards reusable real-time objects. *Annals of Software Engineering: Special Volume on Real-Time Software Engineering*, 7:257–282, 1999.

12. D. Norman. *The Invisible Computer*. The MIT Press, Cambridge, MA, USA, 1998.

13. S. Ren. *An Actor-Based Framework for Real-Time Coordination*. PhD thesis, Department Computer Science. University of Illinois at Urbana-Champaign, 1997. PhD. Thesis.

14. S. Ren, G. Agha, and M. Saito. A modular approach for programming distributed real-time systems. *Journal of Parallel and Distributed Computing*, 36(1):4–12, 1996. Also published in *School on Embedded Systems, European Educational Forum 1996*, pp 52–72.

15. Nalini Venkatasubramanian and Carolyn L. Talcott. Reasoning about meta level activities in open distributed systems. In *Symposium on Principles of Distributed Computing*, pages 144–152, 1995.

Author(s) affiliation:

- **Koushik Sen, and Gul Agha**

  Department of Computer Science
  University of Illinois at Urbana-Champaign
  Email: [ksen,agha]@uiuc.edu

14

Experiments with QoS Driven Learning Packet Networks

Erol Gelenbe
Alfonso Montuori
Arturo Nunez
Ricardo Lent
Zhiguang Xu

Abstract

With the growth of the Internet, in numbers of users, IP (Internet Protocol) addresses, and routers, and the increasing diversity of the types of links that are being used (wired, optical or wireless), the IP based network architecture needs to evolve. Reliability, security, scalability and QoS (Quality-of-Service) have become key issues. This paper discusses a research project which revisits IP routing and suggests a novel packet network architecture called a *"Cognitive Packet Network (CPN)"*, in which intelligent control for routing and flow control are exercised under greater control of the user connections with the help of "smart packets", rather than being completely left to nodes and routing tables. CPN provides a reliable packet network infrastructure which incorporates packet loss and delay directly into user QoS criteria, and uses these and other criteria to conduct routing. We present the QoS based routing algorithm that we have designed and implemented, and report extensive measurements related to network performance, reliability and QoS. Our experiments also show how CPN can address the QoS needs of packetized voice. Other experiments show how CPN can be used to route IP packets which travel to and from a web server and tunnel through a CPN sub-network so as to obtain improved QoS.

14.1 INTRODUCTION

In recent papers [10, 12, 15] we have proposed a new network architecture called "Cognitive Packet Networks (CPN)". CPN makes use of adaptive techniques to seek out routes based on user defined QoS criteria. For instance, packet loss and delay can be used as routing criteria to improve overall reliability for the users of the network, or delay and its variance can be used to find routes which provide the QoS requested by voice packets.

A CPN carries three types of packets: smart packets, dumb packets and acknowledgments (ACK). Smart or cognitive packets route themselves, they learn to avoid link and node failures and congestion and to avoid being lost. They learn from their own observations about the network and from the experience of other packets. They rely minimally on routers. Smart packets use reinforcement learning to discover routes, and the reinforcement learning "reward" function incorporates the QoS requested by a particular user. This reward is the inverse of a QoS "goal" which each user can provide before initiating a connection. When a smart packet arrives to a destination, an acknowledgment (ACK) packet is generated by the destination and the ACK heads back to the source of the smart packet along the inverse route. As it traverses successive routers, it updates mailboxes in the CPN routers; when it reaches the source node it provides source routing information for the dumb packets. Dumb packets of a specific QoS class use successful routes which have been selected in this manner by the smart packets of the same class.

In this paper we first recall the principles underlying the CPN design, and present the CPN test-bed we have implemented. Measurement experiments are reported to illustrate CPN's robustness under link and node failures. Voice over the Internet is a very important application that has stringent QoS requirements; thus it is a very good example for illustrating the capabilities of CPN. We therefore present the manner in which CPN's QoS based routing algorithm can be tailored to voice packets. We discuss the "goal" and "reward" function that needs to be used in this case, and then present experiemntal results we have obtained in CPN for voice communications. Delay, jitter and packet desequencing measurements are reported, and our results are compared with measurements on the same network infrastructure running conventional IP. The QoS improvement offered by CPN over IP is very significant, particularly when the network is heavily loaded. Finally we show how CPN can be used to optimize packet delay for web traffic for conventional IP packets by tunnelling the traffic through a CPN sub-network so as to reduce end-to-end delay.

14.2 COGNITIVE PACKETS AND CPNS

Learning algorithms and adaptation have been suggested for telecommunication systems in the past [5, 8]. However these ideas seldom been exploited in networks because of the lack of a practical framework for adaptive control in packet networks.

A node in the CPN acts as a storage area for packets and mailboxes (MBs). It also stores the code used to route smart packets. It has an input buffer for packets arriving from the input links, a set of mailboxes, and a set of output buffers which are associated with output links. Nodes in a CPN carry out the following functions:

1. A node receives packets via a finite set of ports and stores them in an input buffer. It transmits packets to other nodes via a set of output buffers. Once

a packet is placed in an output buffer, it is transmitted to another destination node with the priority indicated in the output buffer.

2. A node receives information from ACK packets which it stores in MBs. Mailboxes may be reserved for certain classes of packets , or may be specialized by packet classes. For instance, there may be different MBs for packets identified by different Source-Destination (S-D) pairs and different QoS classes.

3. It executes the routing code for each smart packet it receives; as a result of this execution it selects the output link and places the packet in an appropriate output buffer.

14.2.1 Routing Algorithm using Reinforcement Learning

Smart packet routing outlined above is carried out using a reinforcement learning (RL) algorithm [6] based on Random Neural Networks (RNN) [7, 11, 15, 16], and we have also proposed similar algorithms to control autonomous agents in simulation systems [13]. The algorithm code is stored in each router and its parameters are updated by the router. For each successive smart packet, the router computes the appropriate outgoing link based on the outcome of this computation.

A recurrent RNN. with as many "neurons" as there are possible outgoing links, is used in the computation. The weights of the RNN are updated so that decision outcomes are reinforced or weakened depending on how they have contributed to the success of the QoS goal. Earlier simulations of CPN [15], and our current test-bed implementation and experiments, have been used to validate this approach.

The RNN [7] is an analytically tractable spiked random neural network model whose mathematical structure is akin to that of queuing networks. It has "product form" just like many useful queuing network models. The state q_i of the $i-th$ neuron in the network represents the probability that the $i-th$ neuron is excited. Each neuron i is associated with a distinct outgoing link at a node. The q_i, with $1 \leq i \leq n$ satisfy the following system of non-linear equations:

$$q_i = \lambda^+(i)/[r(i) + \lambda^-(i)], \tag{14.1}$$

where

$$\lambda^+(i) = \sum_j q_j w_{ji}^+ + \Lambda_i, \quad \lambda^-(i) = \sum_j q_j w_{ji}^- + \lambda_i, \tag{14.2}$$

where w_{ji}^+ is the rate at which neuron j sends "excitation spikes" to neuron i when j is excited, w_{ji}^- is the rate at which neuron j sends "inhibition spikes" to neuron i when j is excited, and $r(i)$ is the total firing rate from the neuron i. For an n neuron network, the network parameters are these n by n "weight matrices" $\mathbf{W}^+ = \{w^+(i,j)\}$ and $\mathbf{W}^- = \{w^-(i,j)\}$ which need to be "learned" from input data. Various techniques for learning may be applied to the RNN. These include Hebbian learning, backpropagation learning [7], and Reinforcement Learning (RL) [11] which we have used in CPN.

RL is used in CPN as follows. Each node stores a specific RNN for each active source-destination pair, and each QoS class. The number of nodes of the RNN are specific to the router, since (as indicated earlier) each RNN node will represent the decision to choose a given output link for a smart packet. Decisions are taken by

217

selecting the output link j for which the corresponding neuron is the most excited, i.e. $q_i \leq q_j$ for all $i = 1, .. , n$.

Each QoS class for each source-destination pair has a QoS Goal G, which expresses a function to be minimized, e.g., Transit Delay or Probability of Loss, or Jitter, or a weighted combination, and so on. The reward R which is used in the RL algorithm is simply the inverse of the goal: $R = G^{-1}$. Successive measured values of R are denoted by R_l, $l = 1, 2, ..$; These are first used to compute the current value of the decision threshold:

$$T_l = aT_{l-1} + (1 - a)R_l, \tag{14.3}$$

where a is some constant $0 < a < 1$, typically close to 1. Suppose we have now taken the $l - th$ decision which corresponds to neuron j, and that we have measured the $l - th$ reward R_l. We first determine whether the most recent value of the reward is larger than the previous value of the threshold T_{l-1}. If that is the case, then we increase very significantly the excitatory weights going into the neuron that was the previous winner (in order to reward it for its new success), and make a small increase of the inhibitory weights leading to other neurons. If the new reward is not greater than the previous threshold, then we simply increase moderately all excitatory weights leading to all neurons, except for the previous winner, and increase significantly the inhibitory weights leading to the previous winning neuron (in order to punish it for not being very successful this time).

Let us denote by r_i the firing rates of the neurons before the update takes place:

$$r_i = \sum_1^n [w^+(i,m) + w^-(i,m)], \tag{14.4}$$

We first compute T_{l-1} and then update the network weights as follows for all neurons $i \neq j$:

- If $T_{l-1} \leq R_l$
    - $w^+(i,j) \leftarrow w^+(i,j) + R_l,$
    - $w^-(i,k) \leftarrow w^-(i,k) + \frac{R_l}{n-2}$, if $k \neq j$.

- Else
    - $w^+(i,k) \leftarrow w^+(i,k) + \frac{R_l}{n-2}, k \neq j,$
    - $w^-(i,j) \leftarrow w^-(i,j) + R_l.$

Since the relative size of the weights of the RNN, rather than the actual values, determine the state of the neural network, we then re-normalize all the weights by carrying out the following operations. First for each i we compute:

$$r_i^* = \sum_1^n [w^+(i,m) + w^-(i,m)], \tag{14.5}$$

and then re-normalize the weights with:

$$w^+(i,j) \leftarrow w^+(i,j) * \frac{r_i}{r_i^*},$$

218

$$w^-(i,j) \leftarrow w^-(i,j) * \frac{r_i}{r_i^+} .$$

Finally, the probabilities q_i are computed using the non-linear iterations (14.1), (14.2). The largest of the q_i's is again chosen to select the new output link used to send the smart packet forward. This procedure is repeated for each smart packet for each QoS class and each source-destination pair.

14.3 THE CPN TEST-BED: PROTOCOL DESIGN AND SOFTWARE IMPLEMENTATION

In this section we describe the CPN protocol design and the test-bed implementation. The software we have developed has been integrated into the Linux kernel 2.2.x. with minimal changes in the existing networking code, and it is independent of the physical transport layer. The Linux kernel support for low cost PCs and a growing number of platforms, and the freely availability of its source code, makes Linux an attractive system for the development of a project of this nature. The network interface is compatible with the popular BSD4.3 socket layer in Linux. It provides a single application program interface (API) for the programmer to access the CPN protocol.

CPN provides a connectionless service to the application layer, and consists of a set of hosts interconnected by links of some kind, where each host can operate both as an end node of communication and/or as a router. The addressing scheme utilizes a single number of 32 bits to represent the CPN address of each node. All the nodes have been configured to use CPN and IP packets at the same time for comparison purposes. CPs are of variable size and consist basically of three areas: a header, a Cognitive map (CM) and a data portion. Each port of a CPN node uses a 10 Mbps Ethernet link connected with another CPN node. The physical connection between routers uses a crossover twisted pair copper cable. We have tried out various topologies where each CPN router can be connected up to four other routers: Figure 14.1 depicts a typical topology.

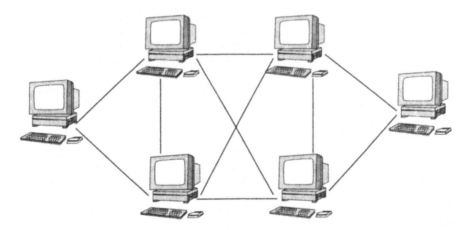

Fig. 14.1 The current test-bed topology

219

14.4 NETWORK MEASUREMENTS

The purpose of the measurements we describe on the test-bed is to evaluate the CPN architecture with respect to its dynamic behavior and ability to adapt to changing network conditions. These conditions include both changes in traffic patterns and possible failures in links. With respect to the network topology described in Figure 14.1, we have conducted measurements for a main flow of traffic from node Node 10 (left) to Node 5 (right). For the first set of experiments, the input rate was fixed to 5 packets second. In several of the experiments we have conducted there is a flow of "obstructing" traffic over and above the main traffic flow from Node 10 to Node 5.

Each time we have added this obstructing traffic it has been on one or more specific links, and at a relatively high traffic rate of 5700 packets per second in each direction (i.e. bi-directional on each selected link).

To simplify matters we have selected the length of the payload for all packets to be of fixed length of 1000 bytes (B). Additionally, there is a CPN header (20 bytes) for each packet. Since we use Ethernet at the transport level we need to add an Ethernet header and tail to each packet. For dumb packets we also have the source route information which acts as the cognitive memory for a total of $16(N-1)$ bytes, where N is the number of nodes in the path. Thus the total packet length is in excess of 1100 bytes. The resulting obstructing traffic is therefore in excess of $5.7MB/sec$ per obstructed link, in each direction.

The $x-axis$ in all the plots refers to successive packets and is scaled in packet counts. In each plot the $y-axis$ presents delay in milliseconds (left) or route number (right). Note that all time values are rounded up to the closest integer number of milliseconds, while the route numberings are indicated in the figure captions. On the plots, an "X" under the $x-axis$ in the route plot indicates an instance of packet loss of either smart or dumb packets.

Figure 14.2 shows a traffic pattern with 20% of SPs and 80% of DPs. Here the network is only carrying the traffic from Node 10 to Node 5 with no interfering traffic. The successive individual delay values and individual routes taken by close to 200 packets is traced packet by packet. Each route label corresponds to some particular sequence of nodes which are traversed.

We observe that routes do change despite the fact that there is no traffic on the network other than the one that is being traced. This is a consequence of sending out a constant proportion of SPs to test alternate routes, resulting in the selection of a new outgoing link at a node if the delay is estimated to be smaller according to the CPN algorithm.

In Figure 14.3 we plot the delays and routes experienced by individual packets on the traffic flow from Node 10 to Node 5. We now have obstructing traffic which perturbs this traffic flow, and we still have 20% of SPs in all traffic flows. The additional traffic is introduced when the packet count reaches 30 on the traffic from Node 10 to Node 5. The obstructing traffic flows on the link from Node 10 to Node 1 . We see that when the obstructing traffic is initialized, the main traffic flow encounters significant delays as well as losses. Then, thanks to the CPN algorithm, the network determines a new route and delays go back to a low level. Further spurious increases in delay occur but are short-lived each time the SPs probe the network for better routes and some losses do occur again.

The network's reaction to link failures was tested and is reported in Figure 14.4. The network has no obstructing traffic during this experiment so that it only carries

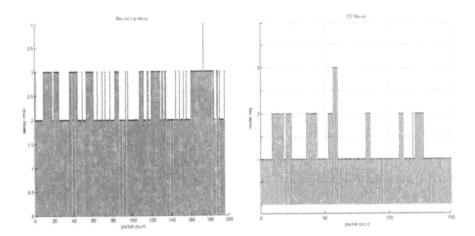

Fig. 14.2 Network with no obstructing traffic.

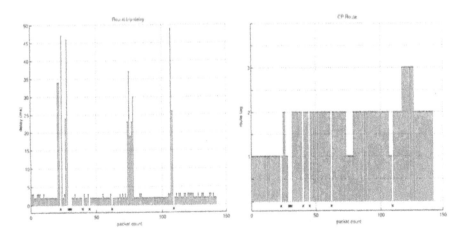

Fig. 14.3 Network with obstructing traffic.

221

traffic from Node 10 to Node 5 with a fixed 20% ratio of SPs. At packet count 40, the link from Node 10 to Node 1 was disconnected and we immediately observed the loss of a few packets. The initially used preferential Route 1 of Nodes 10:1:4:5 was rapidly abandoned by the CPN algorithm, and the system switched to Route 2 which uses Nodes 10:2:1:4:5 = 2. Due to continued searching by SPs, later Route 3 (Nodes 10:2:3:5) and Route 2 are used, but Route 1 is not used again.

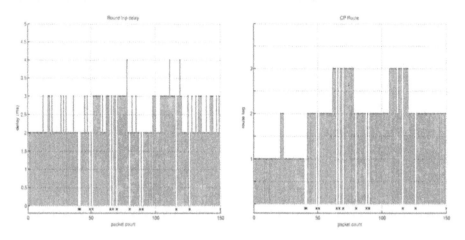

Fig. 14.4 Measurements concerning the network under link failure

In the next set of results, we show the effectiveness of the CPN to adapt to different quality of service requirements, for this, we have introduced jitter control as a new component for the routing decision of the smart packets in addition to the average delay. Along with the main flow of packets we introduce a second flow to produce obstructing traffic to the former. The resulting average delay after the transmission of 200 thousand packets is shown in Figure 14.5. The x-axis represent the inter-packet time for the main flow of packets. There is a significant improvement in the overall standard deviation of the packets when the jitter control is active as shown in Figure 14.6.

Another important aspect that we investigate is the effect of the ratio of SPs on overall performance. We have conducted a set of experiments using a fixed input rate of 2500 packets per second to examine this point and our results are reported in Figure 14.7. The interesting result we observe is that as far as the DPs are concerned, when we have 15% or 20% of SPs we have achieved the major gain in delay and jitter reduction. Going beyond those values does not significantly reduce the delay for DPs.

14.5 TRANSPORT OF VOICE PACKETS IN CPN

Packet based voice transport in the Internet is a critical application which could dramatically increase the traffic carried over IP networks [4]. However, providing desired QoS for voice over IP is a major challenge. The QoS driven nature of CPN offers an opportunity to examine the manner in which network routing can actually serve the needs of voice transport over a packet network, and in this paper we show

222

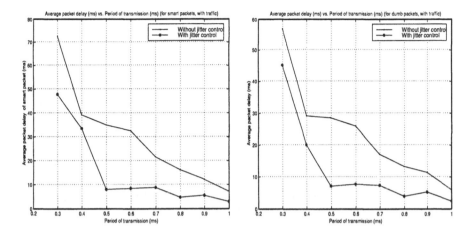

Fig. 14.5 Average round-trip delay for smart (left) and dumb (right) packets under regular routing conditions and routing with jitter control

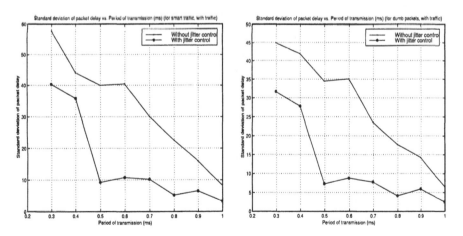

Fig. 14.6 Standard deviation of delay for smart (left) and dumb (right) packets under regular routing conditions and routing with jitter control

223

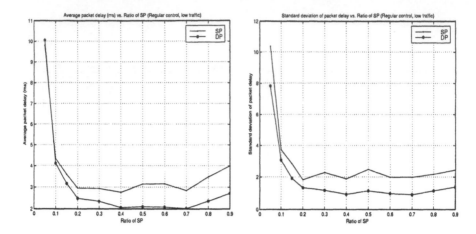

Fig. 14.7 Average round-trip delay (left) and standard deviation of delay for smart (left) and dumb packets as a function of the percentage of smart packets

how CPN routing can be specifically adapted to address the needs of Real-Time voice stream delivery, by incorporating the requirements of RTCP (Real-Time Control Protocol) in the reward function used by the CPN "reinforcement learning" routing algorithm.

Real-Time voice transport introduces tight constraints on QoS with respect to delay, jitter, loss and/or error, due to the limited tolerance of the human listener to both the average delay and the fluctuation of delay [4]. The QoS requirements can be divided into two categories: the end-to-end average network delivery time must be small; the end-to-end variation of the delivery time, as well as losses, must be small. If there is no buffering at the receiver, the overall delay is the sum of three components: the algorithmic delay of the codec, the processing delay of the codec and the network delay. The overall delay should not exceed 200-250 ms, but a delay of 200 ms to 800 ms is conditionally acceptable for a short portion of the conversation, when such delays are rare and far apart. Due to fluctuations of the network delay, buffering is needed at the receiver. If the buffer delay is much greater than mean network delay, the overall delay is dominated buffering period at the receiver and the algorithmic delay and processing delay of the codec.

Unlike the delay, silence periods due to delay fluctuations can compromise seriously the intelligibility of the speech and they must be carefully controlled. The gap structure perceived by the listener will be not only be a function of network load fluctuations but also of the network policy or protocol at the receiver for dealing with these gaps. The RTCP defines a procedure to measure the network delay fluctuation, i.e. the "interarrival jitter". It is the mean deviation (smoothed absolute value) of the difference D in packet spacing at the receiver compared to the sender for a pair of packets. That is: If S_i is the RTCP sender timestamp for packet i, and R_i is the timestamp of the same packet's arrival at destination (measured in RTCP timestamp units), then for two packets i and j, D is:

$$D(i, j) = (R_j - R_i) - (S_j - S_i)$$

224

The interrarrival jitter is calculated continuously as each data packet is received using D measured for successive packet *in order of arrival*:

$$J_i = J_{i-1} + (|D(i-1,i)| - J_{i-1})/16$$

Another important measurement for Real-Time applications is the probability of packet desequencing detected at the receiver, since the receiving application should receive packets in the same order as they were sent. The procedure we use to measure the number of desequenced packets Q is: 1. Set $Q = 0$ at the receiver; 2. At the sender, start sending packets with sequence numbers $1, 2, 3...$; 3. At the receiver, call S_1 the sequence number of the first arrived packet, S_2 the sequence number of the second arrived packet, etc. At the arrival of the j-th packet, increment $Q \leftarrow 1$ if $S_j < max\{S_1, S_2, ...S_{j-1}\}$, else leave Q unchanged; 4. Each time packet loss is detected increment Q by the number of lost packets. Note that in speech transmission, packet losses as high as 50 percent can be tolerated with marginal degradation if such losses occur for very short time intervals (<20 ms).

14.5.1 Reward Function for CPN Routing of Voice

In order to fulfill the requirements of RTCP, we need to incorporate the variance of packet delay in the *Reward* function which is used in our Reinforcement Learning routing algorithm. In CPN, each smart or dumb packet measures the date at which it enters a node, and provides this date information to the ACK packet which heads back from the destination to the source of the corresponding smart or dumb packet. In the sequel we will drop indexing of symbols used with respect to some specific destination; it is assumed that the quantities we discuss are all indexeed separately for each separate destination. Furthermore, these quantities are specific to RTCP QoS class, so that traffic from other QoS classes in the same Cognitive Packet Network may not need them if they do not belong to the RTCP QoS class.

When an ACK (say ACK_i) in CPN reaches some node on its way back to the source, it estimates the forward delay $Delay_i$ from this node to the destination by simply taking the difference between the current time at the node and the time at which the corresponding SP or DP visited the same node, and dividing it by two. This rough estimate is used by CPN to avoid the need for clock synchronization between nodes. The node maintains a smoothed exponential average of these estimated forward delays for each active destination, and we denote it here by $\overline{Delay_i}$. To approximate $D(i,j)$ defined in RTCP we have used:

$$V_i \leftarrow Delay_i - \overline{Delay},$$

and jitter is approximated by

$$J_i \leftarrow J_i + (|V_i| - J_i)/16,$$

and deposited in the node's (router) mailbox, to be used by subsequent smart packets going to the same destination. Now when a smart packet for the RTCP QoS class enters a node, it uses information from the mailbox to computes the reward R for the most recent routing decision:

$$R_i \leftarrow \frac{1}{\overline{Delay_i} + J_i}$$

225

and executes the CPN Reinforcement Learning routing algorithm [16] to select an outgoing link from the node.

14.5.2 Experimental Results

In Figure 14.8 we show the system configuration which has been used while Figure 14.9 shows the CPN test-bed which was used as part of the experiment. The test-bed shown in Figure 14.9 can also be used as an IP network, by simply replacing the CPN protocol software in each of the routers, by the IP protocol stack. Voice traffic originates at machine IP1 as IP packet traffic and its destination is machine IP2. Voice IP traffic from IP1 enters the CPN test-bed at machine CPN10, and leaves the CPN test-bed at machine CPN5 to go to the destination IP2.

Speech is sampled at 8 kHz. The frame length of each packet is 32 ms. The mean value of the bitrate is 12 kbit/s. Voice is digitized by TI's TLC320AD535 16bit data converter on a TMS320C6711 board. The digitized voice is compressed by means of a wavelet packet transform running on the FP C6711 Digital Signal Processor (DSP). The code is written in C and compiled by the Code Compiler Studio optimization software included in the board's toolkit. The compressed voice is sent to the host computer IP1 through the host port interface in real time. The host computer packetizes the compressed voice with the UDP/IP protocol and sends it to the CPN test-bed, where a program called "Tunnelling" running on every CPN interface/edge router does the IP/CPN conversion. At IP2 a jitter buffer of about 200ms is used to avoid data overflow and/or underflow. This buffer permits the reordering of desequenced packets. A server application at IP2 then unpacks speech packets and sends the compressed speech to a DSP board, where it is reconstructed by means of the decoding algorithm and on-board Digital Analog Conversion (DAC). Our experiments compare the QoS for voice traffic over the CPN test-bed, with the same voice traffic carried over an IP network route on exactly the same test-bed.

In the first experiment we send voice from host machine IP1 to machine IP2, connected to the two edge routers CPN10, CPN5, of the CPN network; "silence" is sent from IP2 to IP1 to simulate a two-way conversation.From IP1 to IP2 the voice appears to be transitting through an IP/UDP network, while in fact it either goes through IP or it goes through CPN. Two alternatives (IP and CPN) are compared under identical traffic conditions are created in the network:

- UDP/IP voice packets travel from IP1 to IP2 with IP routing along the path {IP1, CPN10, CPN1, CPN3, CPN5, IP2}.

- UDP voice packets travel from IP1 to CPN10, then are forwarded by CPN QoS driven routing to CPN5, and then become IP packets from CPN5 to IP2IP2.

and In both cases we introduce additional bi-directional data traffic of varying intensity to observe the impact on the voice traffic. In Figure 14.10 (dashed lines) we show how the obstructing data traffic introduced in various network links at different traffic rates. The obstructing traffic bit-rates are set at values greater then 10Mbit/s so as to obtain heavy load conditions. In Figure 14.11 the logarithmic value of the percentage of lost packets is plotted. For IP routing (the upper curve) we observe that the percentage of lost packets grows exponentially as the number of saturated links increases. For QoS driven CPN adaptive routing (the lower two curves), we observe that the percentage of lost packets remains small and quasi-constant. These logrithmic curves provide a clear view of the significant differences in packet loss between CPN and IP.

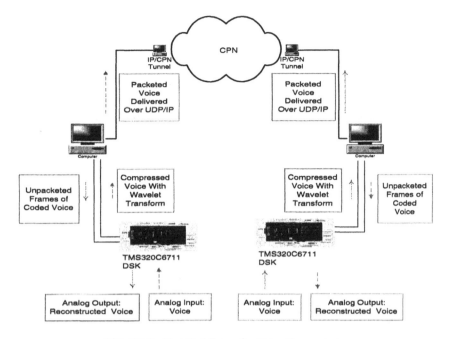

Fig. 14.8   Test-Bed Setup for Voice Transmission

In another experiment, we saturate other links using data traffic (cf. Figure 14.12), so that only one path remains unsaturated. The percentage of lost packets (cd. Table 14.1) remains close to zero, while values of jitter and delay are slightly greater than those from the previous experiment (see Figures 14.9 and 14.10), because CPN now has to select a route with at least 4 hops. The percentage of desequenced packets is under 5%, allowing for fast reordering of voice packets at the receiver.

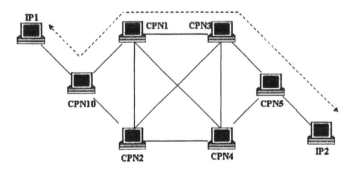

Fig. 14.9   CPN Test-Bed topology and IP (dashed line) route

CPN can in principle allow both SPs and DPs to carry payload, but our standard CPN implementation only places payload in DPs. Thus SPs "take risks" to search for routes providing better QoS, and DPs take advantage of the results of SPs' efforts. The next experiment shows the advantage of this latter approach by showing the

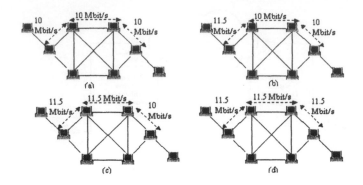

Fig. 14.10 The links marked by dashed lines are progressively saturated

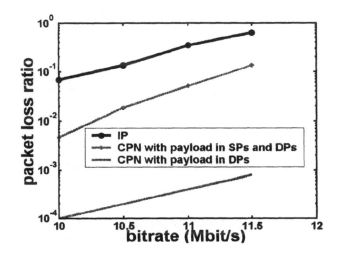

Fig. 14.11 Logarithmic value of the percentage of lost packets plotted as a function of link load

		Delay	Jitter	Desequencing
Fig. 3		9ms	25ms	3.3%
Fig. 5 (a)		14ms	34ms	3.8%
Fig. 5 (b)		14ms	36ms	4.0%
Fig. 5 (c)		13ms	32ms	4.3%
Fig. 5 (d)		16ms	36ms	4.1%

Table 14.1 CPN QoS is robust to increases in obstructing traffic

performance difference between the case where both SPs and DPs carry payload, and

228

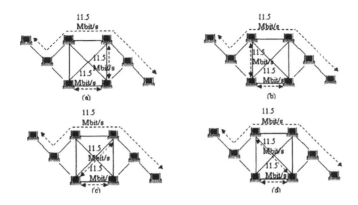

Fig. 14.12 More links become saturated

the case where only DPs carry payload. In the following figures the x-axis represents the inter-packet arrival time of the flow of packets from node CPN10 to node CPN5 (refer to Figure 14.9). There is also a flow of obstructing traffic from CPN10 to CPN5 as illustrated in Figure 14.10. In the lower curves of Figures 14.13 and 14.14, we see that that if only DPs carry payload, both average delay, its standard deviation (jitter), and the packet desequencing probability remain (relatively) constant when input traffic rate varies widely. In contrast, when both SPs and DPs are allowed to carry payload (curves above), both delay, jitter and the packet desequencing probability increase dramatically with increasing input rate.

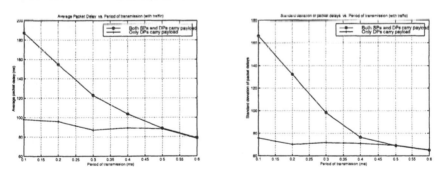

Fig. 14.13 Delay (left) and Jitter (right) measurements for Voice over CPN. Only DPs carry payload (curves below), and both DPs and SPs carry payload (curves above)

14.6 USING CPN TO CONTROL IP TRAFFIC

An interesting application of CPN technology is as a "control network" which can be used to direct traffic in the Internet. The idea is to have a CPN network having multiple output ports into the Internet. Assume a user wishes to direct his IP packets to some destination D which lies within the Internet, but with the desire to obtain a

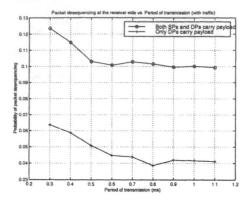

Fig. 14.14 Probability that a packet arrives out of sequence at the receiver: Only DPs carry payload (curve below), and both DPs and SPs carry payload (curve above)

certain type of QoS – say minimum delay to destination. The user then directs his packets first into a CPN network which has multipleoutput ports into the Internet (e.g. via different ISP providers). The CPN network tunnels these IP packets, disguising them as CPN packets which have the destination D which is represented inside CPN as any output port in the Internet. However, CPN's QoS driven routing algorithm will direct the IP packets dynamically to that output port which offers the best QoS according to the end user's criterion, e.g. minimum delay. Then from the output port the packets once again become IP packets and travel to destination D in the usual way.

Figure 14.15 shows the experimental set-up we have used to demonstrate this capability. A web server's output port is connected to CPN, but this web server receives requests from users via the Internet using IP. In response to a web access, the web server sends IP packets into the CPN network where they tunnel to two possible output ports (A30 and A40) from which they will go back into the Internet and then head towards the destination over the Internet. The selection of the A30 or A40 port is made by CPN based on the user's or the web server's QoS goal. In Figure 14.16 we show experimental results obtained with this scheme. The $y - axis$ shows the percentage of total outgoing traffic which uses the A30 output port into the Internet. The $x - axis$ provides the data points which were used, which are the difference in delay δ between the A40 output port and the A30 output port. Thus when δ is positive, this means that A30 provides lower delay, and as we see from the figure it results in a greater percentage of traffic taking the A30 output port. The opposite effect is also observed at the left hand side of the figure. This shows that CPN can be effectively used to control the flow of traffic *beyond* CPN itself and into the Internet.

14.7 CONCLUSIONS

Cognitive Packet Networks (CPN) are a new packet network paradigm which address some of the needs of global networking. CPN simplifies router architecture by transfering the control of QoS based best-effort routing to the connections using smart packets. Routing tables are replaced by reinforcement algorithm based routing

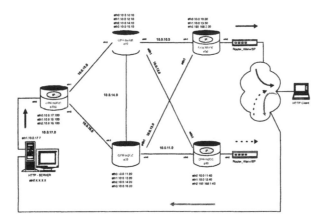

Fig. 14.15 Experimental set-up for CPN control of Web Server traffic

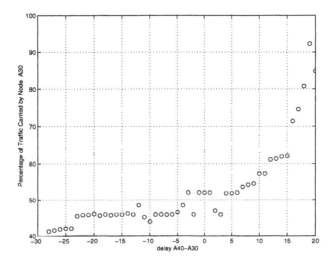

Fig. 14.16 Web traffic flow to destination via alternate ports A30 and A40 as a function of the difference in delay from the output ports to destination

functions. A CPN carries three distinct types of packets: Smart or cognitive packets which search for routes based on a QoS driven reinforcement learning algorithm, ACK packets which bring back route information and measurement data from successful smart packets, and Dumb packets which do source routing.

In this paper we have summarized the basic principles of CPN. Then we have described the Reinforcement Learning (RL) algorithm which taylors the specific routing algorithm to the QoS needs of a class of packets. In order to evaluate the effectiveness of smart packets, we have derived analytical results for their best and worst-case performance. We have then described in some detail the design and implementation of our current test-bed network which uses ordinary PC-based workstations as routers.

231

CPN software has been implemented in a Linux environment and is portable to a wide range of platforms.

We have provided measurement data on the test-bed to illustrate the capacity of CPN to adapt to changes in traffic load and to failures of links. Some of the measurements we report present traces of short sequences (under 200 packets) of packet transmissions showing individual packet delays, paths taken by the packets and individual packet losses due to link failures or buffer overflows. We also report long-term measurements covering hundreds of thousands of packet transmissions. These long term measurements report average packet delays for Smart and Dumb packets with and without obstructing traffic. They also show how different QoS goals (e.g. Delay or Jitter) impact performance. We also measure packet loss rates, rates of packet desequencing, Measurements are also reported to evaluate the impact of the ratio of Smart packets to total packets, on the end-to-end delay experienced by all of the packets. We see that a 10-20% ratio of Smart packets is best for the performance of Dumb packets. In heavy traffic conditions, if the ratio of Smart packets becomes too high, the Dumb packet delay can become very significant. Finally, we have shown how the QoS driven routing algorithm used in our novel Cognitive Packet Network can be specifically designed to address the needs of store-and-forward routed packetized voice and also to optimize end-to-end delay for web traffic.

Our ongoing work includes the deployment of a large test-bed, the inclusion of wireless links, and the design of single-card routers leading to very low cost routing technologies which use CPN.

REFERENCES

1. E. Gelenbe "Probabilistic automata with structural restrictions", *Proc. SWAT 1969 (IEEE Symp. on Switching and Automata Theory)*, also appeared as "On languages defined by linear probabilistic automata", *Information and Control*, Vol. 18, February 1971.

2. E. Gelenbe "A realizable model for stochastic sequential machines", *IEEE Trans. Computers*, Vol. C-20, No. 2, pp. 199-204, February 1971.

3. R. Viswanathan and K.S. Narendra "Comparison of expedient and optimal reinforcement schemes for learning systems", *J. Cybernetics*, Vol. 2, pp 21-37, 1972.

4. D. Minoli, E. Minoli, *"Delivering Voice over IP Network"*, John Wiley & Sons, New York, 1998.

5. K.S. Narendra and P. Mars, "The use of learning algorithms in telephone traffic routing - a methodology", *Automatica*, Vol. 19, pp. 495-502, 1983.

6. R.S. Sutton "Learning to predict the methods of temporal difference", *Machine Learning*, Vol. 3, pp. 9-44, 1988.

7. E. Gelenbe (1993) "Learning in the recurrent random neural network", *Neural Computation*, Vol. 5, No. 1, pp. 154-164, 1993.

8. P. Mars, J.R. Chen, and R. Nambiar, "Learning Algorithms: Theory and Applications in Signal Processing, Control and Communications", CRC Press, Boca Raton, 1996.

9. E. Gelenbe, Zhi-Hong Mao, Y. Da-Li (1999) "Function approximation with spiked random networks" *IEEE Trans. on Neural Networks*, Vol. 10, No. 1, pp. 3–9, 1999.

10. E. Gelenbe, E. Seref, Z. Xu "Towards networks with intelligent packets", *Proc. IEEE-ICTAI Conference on Tools for Artificial Intelligence*, Chicago, November 9-11, 1999.

11. U. Halici, "Reinforcement learning with internal expectation for the random neural network" *European Journal of Operations Research*, Vol. 126, no. 2, pp. 288-307, 2000.

12. E. Gelenbe, R. Lent, Z. Xu "Towards networks with cognitive packets," Opening Key-Note Paper, *Proc. IEEE MASCOTS Conference*, ISBN 0-7695-0728-X, pp. 3-12, San Francisco, CA, Aug. 29-Sep. 1, 2000.

13. E. Gelenbe, E. Şeref, Z. Xu, "Simulation with learning agents", *Proceedings of the IEEE*, Vol. 89 (2), pp. 148–157, Feb. 2001.

14. E. Pasero, A. Montuori, "Wavelet Based Wideband Speech Coding on the TMS320C67 for Real Time Transmission", *Proceedings of IEEE MTAC 2001, Multimedia Technology and Applications Conference*, Irvine, 8-11 Nov. 2001.

15. E. Gelenbe, R. Lent, Z. Xu, "Towards networks with cognitive packets", Opening Invited Paper, *International Conference on Performance and QoS of Next Generation Networking*, Nagoya, Japan, November 2000, in K. Goto, T. Hasegawa, H. Takagi and Y. Takahashi (eds), "Performance and QoS of next Generation Networking", Springer Verlag, London, 2001.

16. E. Gelenbe, R. Lent, and Z. Xu, "Design and performance of cognitive packet networks," *Performance Evaluation*, Vol. 46 (2001), pp. 155-176.

Author(s) affiliation:

- **Erol Gelenbe, Alfonso Montuori, Arturo Nunez, Ricardo Lent, Zhiguang Xu**

  School of Electrical Engineering and Computer Science
  University of Central Florida
  Orlando, FL 32816, USA
  Email: [erol,alfonso,anunez,rlent,zgxu]@cs.ucf.edu

15

A Perspective on a Mobile Ad Hoc Network: The IRULAN Project

Philippe Jacquet

Abstract

We present the mobile ad hoc networks that allow mobile nodes to communicate without pre-existing infrastructures. These networks can be used for a high speed wireless access to internet in urban area (IRULAN project). We introduce the protocol OLSR whose optimization with respect to link state routing allows the use in extreme mobility. Extensions such as multicast and power saving are described.

15.1 INTRODUCTION

Mobile ad hoc networks have experienced a growing interest this recent years. Their key concept is based on mobile nodes communicating by radio without fixed infrastructure. Data transfers between mutually out of range pairs of node are achieved via multi-hop internal routing. Applications for such networks are natural in military operations and emergency situation. During 9/11 attack, 15 base stations have been instantaneously knocked out, leaving the mobile network in such a disorganized state that it was impossible to make a call in the area for several days, due to non interoperable and re-routable communication processes . As members of the Wireless Emergency Rescue Team (WERT) the telecom operators sent volunteers hiking on the rubble with cable and portable base stations on their back in order to collect emergency calls from survivors.

In a more *peaceful perspective*, Ad hoc networks are interesting alternative to 3G+ high speed internet mobile access internet. Indeed ad hoc networks are more and more considered as the foundation ambient network for wireless internet access: they offer more bandwidth, they are less costly than CDMA systems and more easy to install. Recently an expert of a Finnish startup company remarked that with only the cost of one 3G licence in Germany one could have installed WiFi access points every 80 meters on the whole German territory. The installed park of WiFi and Bluetooth technology is already enormous (there are several millions 802.11 chip set sold in the world and prices are dropping below $100 each). They can dynamically be deployed with evolutionary extensions based on low power device which can be installed anywhere. This has to be compared with telecom operators effort to convince owner to accept on their roof expensive base stations, whose transmission power should have non negligible impact on the health for people living or working in the building.

We have proposed the wireless structure called Internet Radio Urban Local Ambient Network (IRULAN), to open the perspective of wireless internet access based on ad hoc networks. This article will describe such structure. First we give a short review about wireless technology and its performance and limitations. Second we describe the IRULAN project. Third, we compare with other general wireless architecture such as base station based infrastructure. Fourth We go into more detail on ad hoc architecture and routing. Fifth We look on performance on routing protocols such as the Optimized Link State Routing protocol. In a sixth section we will discuss on services and extension specific to ad hoc architecture.

15.2   WIRELESS TECHNOLOGY SURVEY

When we look to the performance of present radio devices, it clearly comes that one must compromise between range and capacity. The mobile phone technology such as GSM enjoy large range (up to several kilometers) but the capacity is relatively low (lower than 100 kbps). On the other hand IEEE 802.11b offers around 10 Mbps but limited to several tens of meters. The reason of this situation is in three letters: DSP. The digital signal processing is the main factor that limit the range when the channel rate increases. Multipath and echoes on obstacle make that he signal received is the combination of various shift in time and in phase of the same signal. To recover the signal one needs to equalize the channel, i.e. to inverse linear systems made of signal sampled at the rate of the channel. One calls the ratio between echo shift and symbol duration, the symbol inter-correlation factor. the larger the symbol inter-correlation factor, the more sampled signals are needed, increasing the cost of DSP per symbol. Since for a same echo drift, the symbol inter-correlation factor increase linearly with the symbol rate, it comes that for a same echo drift the cost of DSP per time unit increases as the square of the symbol rate, making it the main limitation factor of terrestrial networks. Notice that satellite or directive line of sight radio channel are not subject by these limitations since their beam are propagated in free space without obstacle.

Figure 15.1 displays the performance of existing wireless products in terms of capacity and range. We notice that most of existing products stand on the same line between IEEE 802.11a and GSM. The reason comes from the fact that they all represent the same generation of wireless product. The resemblance is such that two very different products such as GSM portable phone and Wavelan wireless LAN chipset present the same DSP power consumption: 40 mA. The Bluetooth device

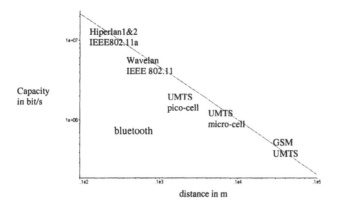

Fig. 15.1 Range-capacity chart of wireless products

presents significantly lower performance than the other because of the goal to achieve smaller power consumption.

This figure clearly illustrates the need to use internal routing and short range for high speed wireless networks. To make a virtue from a vice, it is clear that short range does not require high transmission power (the echos will be amplified as well): therefore high speed wireless device are also low power radiating device, significantly reducing the impact on health. This has also another advantage for military purpose: the low radiation makes the ad hoc network a "stealth" network difficult to detect by an enemy.

15.3 THE IRULAN PROJECT

The IRULAN project consists into covering a city or the center of a city with an ad hoc network. Contrary to other wireless experiment in cities where access points are added in order to cover the maximum area without too much concertation, the IRULAN wireless devices will cooperate in order to have a balanced routing policy between them. The ad hoc network will be based on multiple interface mini-routers dispatched on the area to cover. The aim of the mini-router is to route high speed data between mobile nodes and internet access points. In the beginning of the deployment there maybe only few internet access points connected to a cabled network, increasing the number of packet retransmissions in the air (see Figure 15.2). When the network becomes popular and needs more bandwidth, one proceeds to a densification of the access points decreasing the number of packet retransmissions between the mobile and the access point. Nevertheless some areas such as parks, monument, private lots won't be equipped with access points and will still need internal routing in air.

Thanks to a dynamic network management mobile nodes are tracked by the protocol and packet rerouting will follow optimal path in order to achieve the best performance. The proactive ad hoc routing protocol OLSR[2] is particularly well suited for this task. Mobility is an even more general concept. In a noisy environment the propagation medium can vary much faster than the proper motion of the mobile. For example a large truck can mask the direct path to a router, a gate can be closed or

237

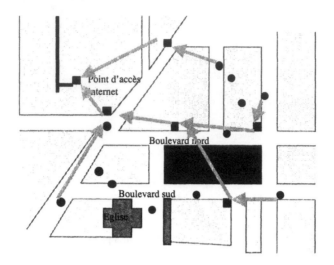

Fig. 15.2 Urban use of ad hoc network

a location can be locally congested, urging a fast data rerouting process. In case of power break or in a disaster, some routers can be temporary knocked out, the network should automatically reconfigure to face the new situation in an optimal way.

15.4 BASIC WIRELESS ARCHITECTURES

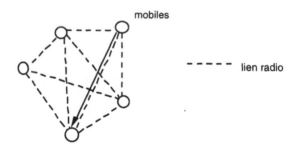

Fig. 15.3 Full connectivity architecture

There are three main basic wireless architecture. Historically the first one is the full connectivity network: every node is a powerful transmitter with low channel rate which can be heared by every other nodes (e.g. the old wireless Morse). Therefore all nodes can be directly in communication with any other nodes as depicted in Figure 15.3. With existing current technology, it will be equivalent to have anyone in the same room.

The most popular architecture is nowadays the base station architecture (Figure 15.4). It has been popularized by the mobile phone concept under GSM. Every mobile

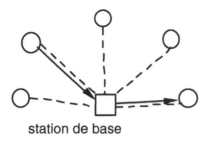

station de base

Fig. 15.4 Base station architecture

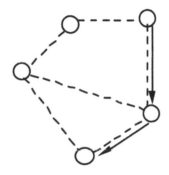

Fig. 15.5 Internal routing architecture

node communicate through a base station located at the center of a cell. Therefore it is important that every node in the cell be in range of the base station. This condition is important but anyhow is less restrictive than the condition of the full connected network where all mobile must be in range of each other. When a node leaves a cell, it changes it disconnect from its base station and connect to the base station of the cell it enters (handover). The base station must be located in a central position with high elevation. The base station between cell are connected via a wired network, directly connected to the public phone network. In the GSM there is a frequency division between cells, such that adjacent cells always operate distinct frequencies. However the frequency division is not needed in all base station architecture, for example in IEEE 802.11 base station infrastructure, adjacent base stations can operate on same channel, or overlapping channels.

The internal routing architecture is the most general architecture one can imagine. The nodes manage among themselves the routing of data between nodes which are not in mutual range (Figure 15.5). This architecture is more reliable than the base station architecture since the requirement for a correct functioning is limited to the transitive connectivity of the graph of the network. In the base station architecture there is the requirement that at least one node (the base station) has a valid link to all other nodes in the network, which much more restrictive. In particular an internal routing architecture is not limited in physical deployment since the number of hop can be as large as one wants. The base station architecture needs to be deployed within a diameter of two physical radio range. The full connectivity graph must fit in single radio range diameter.

15.5   MOBILE AD HOC NETWORK ARCHITECTURE

The ad hoc architecture is the internal routing architecture where the nodes are mobile and have unpredictable motion. Therefore the topology of the network, i.e the graph made by the valid links of the pair of nodes in mutual range, is subject to unpredictable change, maybe experience split and merge (see Figure 15.6).

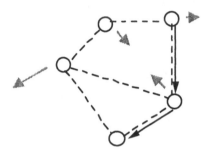

Fig. 15.6   Mobile ad hoc architecture

Since the nodes are mobile, there is a need to rely on a dynamic topology management. This concept is well known since ARPANET when nodes update their routing table very hour, but the nature of radio networks complicates considerably the things. The first and main problem is in the fact that ad hoc networks are much more mobile than their wired counterpart but with much less bandwidth. In general wireless device work around several Mbps, new generations will offer up to 100 Mbps but with a shorter range. This must be compared with the tens of Gbps and Tbps that are common on wired networks. In consequence the traffic of control may swamp the capacity of the network. In a wired network a periodic update of routing table every hour suffices to catch the non frequent changes of topology. The radio range compared to potential motion speed of the mobile nodes (walking people, vehicles) calls for a refresh period of routing information of the order of the second. Indeed a vehicle typically travels a radio range in less than 30 seconds at 30 kmph. The refresh rate should be increased if one increases the channel rate since the radio range decreases when the channel rate increases. Since the radio range decreases faster than the inverse of the channel rate, increasing the refresh frequency proportionaly to the channel rate is not a solution. The solution is to optimize the routing algorithm in order to face these critical conditions.

Another problem is that one expect the mobile ad hoc networks to be rather large with high density. One talks about $1,000-10,000$ nodes. In this case the node degree, I.e the number of neighbor nodes maybe very large $100-1,000$, much more than one could expect on a wired network where routers have limited interface numbers. In this case the overhead of classical routing protocol such as OSPF can be overwhelming. Let consider a network with N nodes (e.g $N = 10,000$), each one having in average number M of neighbor (e.g $M = 1,000$). During every refresh period, every node generate one LSA message, every LSA message contains in average M node ID. Every LSA message is retransmitted $N * M$ times by the nodes (one transmission per link). Therefore the control traffic wastes a fraction of the network capacity equivalent of $(N * M)^2$ ID per refresh period, which is tremendous and does not fit the conditions of operations of an ad hoc network.

The aim of routing in ad hoc network is to define protocols which fit to these severe conditions. The Internet Engineering Task Force (IETF) has created a working group Mobile Ad hoc NETwork (MANET) whose aim is to standardize ad hoc routing protocol [1]. There are two classes of routing protocol, the proactive class and the reactive class.

The proactive class contains the protocols with which the nodes maintain routing information for all possible destinations in the network, and use periodic messages to update them. The Optimized Link State Routing (OLSR) is the example of such protocols. Proactive protocols are good for uniform traffic with many simultaneous connection with many different destination coexist. The proactive protocols must drastically optimize their control traffic overhead in order to fit the conditions of ad hoc network.

The reactive is more revolutionary in its approach. It contains the protocols with which the nodes maintain routing information only on the destinations for which they have traffic toward them, and open route only on demand. Reactive protocol are good for network with a small numbers of simultaneous connection. Since the route opening and repair frequently relye on full flooding, the reactive protocols generate lot of overhead when mobility and link failure increase. Reactive protocols have also difficulties to adapt to flow controlled traffic like TCP because route initialization happens in severe condition where a large number of nodes overload the channel during the route discovery.

15.6 OLSR PROTOCOL, PERFORMANCE OF AD HOC ROUTING

OLSR [2] is derived from link state OSPF protocol [3]. We have seen that OSPF generates a much too large control traffic overhead. One easy way to reduce the control traffic of OSPF consists to replace the unicast flooding with the broadcast flooding. In the unicast flooding every node retransmit a copy of the packet to be flooded to any of its neighbor, leading to $N * M$ retransmission. In the broadcast flooding, one take advantage of the broadcast nature of the radio interface: every node broadcast only once a single copy of the packet to be flooded toward all its neighbor. Therefore the broadcast flooding leads to only N retransmission, which is still too much.

The protocol OLSR introduces a more drastic optimization by using the concept of MultiPoint Relay (MPR) in order to reduce the flooding overhead. In the MPR flooding only a selected subset of nodes retransmit the packet to be flooded. Every node selects its MPR set among its neighbor. The MPR set of a node must cover the 2-hop neighbor which is easy to check since hello messages generated by neighbor contain their neighbor list (see Figure 15.7). A MPR selector of a node is neighbor which has selected this node as MPR. The rule of MPR flooding is that a node forward a packet to be flooded only if it received the first copy from a MPR selector (see Figure 15.8).

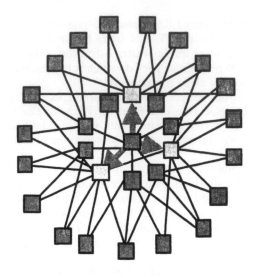

Fig. 15.7 MPR set

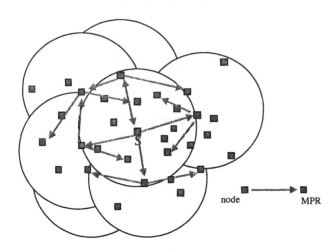

node ──────► MPR

Fig. 15.8 MPR flooding

The smaller is the MPR set per node, the smaller will be the number of retransmissions in an MPR flooding [5]. The choice of the optimal MPR set for a node is an NP problem but there are simple heuristic. For example the greedy heuristic [5] has a complexity proportional to the number of link between the neighbors and the 2-hop neighbors. The greedy heuristic is easy to implement and is proven to select an MPR set close to the optimal one within a factor $logM$.

The second drastic optimization in OLSR compared to OSPF is in the content of the LSA messages. In OSPF the LSA generated by a given node contains the list of neighbors of this node. The LSA message of a node operating OLSR (called for instance Topology Control (TC) messages) restricts its content to the list of MPR selector nodes. Therefore if D_M and R_N respectively denote the average size of the MPR set per node, and the average number of retransmissions in an MPR flooding, then the cost of TC transmission per refresh interval is $R_N * D_M * N$. Since we expect $R_N << N$ and $D_M << M$, the gain over OSPF is tremendous (see Figure 15.9).

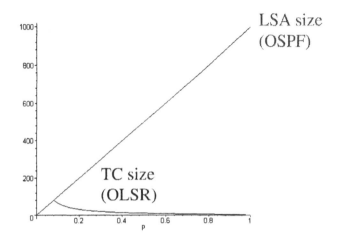

Fig. 15.9 Control message size in random graph model ($1,000$ nodes, p variable)

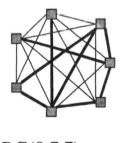

RG(0.7,7)

Fig. 15.10 A random graph with 7 nodes

We have D_M of order of $log M$, 2 and $M^{1/3}$ depending on network models. We also have R_N of order $log N$, N/M, $N : M^{2/3}$.

The known graph by a node operating OLSR is therefore limited to its adjacent links and the MPR links advertized in TC messages. To compute a route to a given destination, the node use a shortest path Dijkstra algorithm on its known graph. One could think that a partial topology knowledge would lead to sub-optimal route compare to the full topology knowledge leading to a waste of bandwidth due to unnecessary

243

retransmissions of data packets. This is true in general but not with OLSR, the optimal path in the known graph are also optimal in full graph. This property comes the 2-hop neighborhood coverage of the MPR set. Therefore there is no compromise between partial topology knowledge and route optimality in OLSR [4].

There are two kind of basic models for ad hoc network depending on the utilization:
1. The random graph model for indoor utilization;
2. The distance graph model for the outdoor utilization.

In indoor conditions the presence or absence of obstacles such as furnitures, walls, is the main cause of link existence between two nodes. Therefore considering random obstacle disposition and random node location, the random graph model is relevant (see Figure 15.10). Of course it is not completely realistic since in the random graph model links exist independently of the other links, but this is a good starting point to get qualitative results as those displayed in figures.

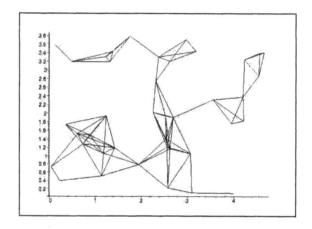

Fig. 15.11 A distance graph with 40 nodes

In outdoor conditions the distance between the nodes and the path loss is the main cause of link existence. The distance graph consists to dispatch randomly nodes on an area (for example a rectangle as in Figure 15.11), and to link the pair of nodes which are at distance smaller than the radio range (in general set as the distance unit). We can imagine the distance graph in dimension one. In this case this models a network of vehicle located on a road. In dimension 2, it will model a battlefield with moving units. In dimension 3 it will model an air traffic monitoring system.

The distance graph model is not completely realistic since random fading may cumulate with distance attenuation to impact the quality of a link. A more realistic model would be a bribing between random graph model and distance graph model. In any case the modelization of ad hoc networks is a very difficult problem since the quality of a given link may also be impacted by the actual traffic on other links, not

244

necessarily adjacent (hidden nodes, exposed nodes) and the impact of the MAC and physical layer should not be ignored.

Mobility can be easily modeled in both models. In the random graph model, one affect a life time to link quality: a link keeps its status unchanged during a random time (preferably exponentially distributed). More generally the status of a link can follow a continuous time Markov process. In the distance graph one affect a random motion evolution to each node, following a Markov process: during a random time the node stay on its current position, then it selects a random speed and keep moving at constant speed during a random time, before stopping again and restarting the process.

The performance parameters that characterize an ad hoc network are the same as for the other networks excepted the first one: reliability. We list the following parameters
- Reliability: the probability that the network can connect any pair of nodes;
- Control overhead;
- Packet delivery ratio; packet delay.

15.7 EXTENSION AND SERVICES ON AD HOC NETWORKS

15.7.1 Power Saving Extension

As we mentioned in introduction, the power consumption is a critical problem. If nodes were consuming power only in packet transmission, then problem will reduce simply to limiting packet generation and forwarding on power critical nodes. For example the power critical nodes will be identified as non-forwarder and never be selected as relay by the other nodes. Unfortunately as we have seen in introduction the DSP is a major source of power consumption. In IEEE 802.11b the node consumes an equivalent quantity of energy while decoding an incoming packet as in transmitting a packet. Therefore there is a need not only to mute a power critical node but also to switch off their reception device. The latter is very much less easy to implement because one must store somewhere the packets addressed to the node when it is in a sleep period.

IEEE 802.11 specifies power saving by allowing power saving nodes to periodically switch between sleep periods and wake periods, e.g. 10 times per second. If sleep period are 10 times longer than wake period then the autonomy of the wireless device is multiplied by ten. In IEEE 802.11 the base station plays the role of power supporter, i.e it stores the packet addressed to the sleeping node waiting for its next wake period to deliver them.

This process is adaptable to ad hoc architecture. The power saver node elects some neighbors as power supporters as in Figure 15.12. The latter will receives the packets en lieu et place of the power saver node and deliver them when appropriate. In OLSR the power supporter nodes plays the role of the MPR nodes of the power saver node since all traffics toward this node go naturally through its MPR selector.

When the network is essentially made of portable device with limited power supply, an upper layer protocol is needed to arbitrate among the nodes those allowed to sleep mode in order to keep the network connected. The status of sleeping node must circulate enough frequently between node in order to allow them to save their battery.

Power saving node   Power supporter MPR

Fig. 15.12   Power saving versus power supporter

An interesting performance measure is the time before the first device exhausts its battery.

15.7.2   Multicast Extension

Since the channel bandwidth is scarce, all means to reduce the overload utilization of the medium are welcome. The multicast data delivery is among those means. When several user want to access to the same data, i.e a journal or a TV channel, classic servers allocate as many unicast flow there are users. Several video streams can kill a wireless network. Therefore it is mostly appropriate to use multicast relaying of data, and to take advantage of the broadcast nature of radio transmissions. In a multicast relaying, packet transmitted in multipoint mode, which allows the server to transmits the data once in a single flow instead of, say 100, parallel flows.

Since the network is multi-hop and the group of multicast receiver could be spread in several location, the multicast flow will be relayed and forked (relayed by different routers in different path) in order to reach all the intended destinations. The union of the relay path is called the multicast tree and is the backbone of the multicast transmissions from the source to the multicast group of receivers (see Figure 15.13). The optimization of the multicast tree, i.e the minimization of the number of its internal routers is an important factor, but not as important as its reactiveness to topology changes and its reliability. Indeed the multipoint mode is not acknowledged and is thus prone to packet loss. If the multicast tree presents some redundancy it may reduce the packet loss.

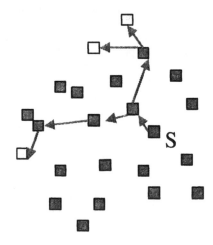

Fig. 15.13   A multicast tree in ad hoc network

246

If the existing redundancy of the tree is not sufficient to keep the packet loss rate below some targeted threshold, there exists multipoint ARQ protocols that gather information from destinations to source in order to trigger packet retransmissions.

Since OLSR is a proactive protocol and provides a suitable topology knowledge, it is particularly well fitted for the building up and failure repair of multicast tree. A multicast extension exists called Multicast Optimized Link State Routing (MOLSR).

15.7.3 Quality of Service,(QoS) Routing

The multimedia services are resource consuming, it is important to have a good policy to avoid that too many multimedia services (even in multicast mode) swamped the radio channel. An admission control based on quality of service can regulate the access to the network. The quality of service management in ad hoc network is more complicated than in regular since the packets are likely to be retransmitted several time due to multihop routing. Consequently a multimedia flow can consume resource proportionally to its number of hops. The challenge here is that the number of hops can vary during the connection, when the user moves. The QoS should be able to track the changes in path length or to anticipate these variation according to the user privileges. Furthermore, some areas of the network could be less congested than other areas changing the condition of admission control when the user moves from one area to another one.

15.7.4 Localization and Services

In a base station architecture the localization is greatly facilitated by the fact that base station are fixed. In an ad hoc network this is less easy. There are several ways to get positioning. In outdoor, one can localize all moving nodes via successive triangulation (using signal strength) from a group of fixed nodes, or nodes which can rely on an external positioning system (such as GPS). In indoor, the triangulation won't perform as well because of too strong random signal fading due to obstacle (furniture, walls). In this case triangulation can be replaced by a signal matrix database, which basically list all the signal strength when transmitter and receiver nodes are in all possible different rooms. The database should contain some redundancy to keep track of event that can affect signal strength (e.g. doors open or closed).

Services that can take advantage of localization are numerous. One can imagine a service which identifies the closest printer for an user with a document to print. Service people can bring food to people located in a room just on demand on a palmtop. In a city environment, taxi can locate their customers, party can meet, etc.

REFERENCES

1. S. Corson, J. Maker, Mobile ad hoc networking and the IETF, 1998

2. draft olsr in the IETF manet working group http://www.ietf.org/internet-drafts/draft-ietf-manet-olsr-06.txt, 2002.

3. OSPF specification, RFC 1131, IETF.

4. T. Clausen, G. Hansen, L. Christensen and G. Behrman. The Optimized Link State Routing protocol, evaluation through experiments and simulation. IEEE Symposium on Wireless Personal mobile Communications, 2001.

5. A. Laouiti, A. Qayyum, L. Viennot, Multipoint relaying: an efficient technique for flooding in mobile wireless networks. 35th Hawaii International Conference on System Sciences (HICSS'2001), 2001.

Author(s) affiliation:

- **Philippe Jacquet**

  INRIA, Unité de recherche de Rocquencourt
  Le Chesnay Cedex - France
  Email: Philippe.Jacquet@inria.fr

16

Semi-Discrete Matrix Transforms (SDD) for Image and Video Compression

Sacha Zyto
Ananth Grama
Wojciech Szpankowski

Abstract

A wide variety of matrix transforms have been used for compression of image and video data. Transforms have also been used for motion estimation, quantization, and image classification. One such transform, that is frequently used to reduce dimensionality of underlying data is the singular-value decomposition (SVD), which relies on low rank approximations for computational and storage efficiency. In this paper, we describe the performance and use of a variant of SVD in image and video compression. This variant, first proposed by Peleg and O'Leary [5], called semi-discrete decomposition (SDD), restricts the elements of the outer product vectors to 0/1/-1. Thus approximations of much higher rank can be stored for the same amount of storage. We demonstrate the superiority of SDD over SVD for a variety of compression schemes. We also show that DCT-based compression is still superior to SDD-based schemes in terms of compression ratios. However, reconstruction from compressed vectors is much faster in the case of SDD compressed images. Furthermore, SDD can be used for fast and accurate pattern matching and motion estimation; thus presenting excellent opportunities for improved compression.

16.1 INTRODUCTION

An image can be viewed as a matrix A of size $m \times n$, where component A_{ij} represents the grey-level intensity at the point (i, j) (generalization of this representation to color images can be easily derived). Rather than storing this entire matrix, lossy compression techniques approximate it using elements that require less storage space. The matrix A can be written as the sum of $r = \min(m, n)$ outer products: $A = USV^T$, with both matrix U and V being orthogonal, and $S = diag(s_1, s_2, \ldots, s_d)$, such as $s_1 \geq s_2 \geq \ldots \geq s_d$. This is the Singular Value Decomposition (SVD) [3] of the matrix A. The SVD decomposition truncated at level k, with $k \leq r$ often provides a fast converging approximation of A. However, even if there are fewer terms (lower rank), each element of the outer product vectors still needs to be stored at least as a byte. (Experiments show that reducing the element storage results in unacceptable deterioration in image quality). On the other hand, SVD provides a well understood basis for approximating A, with guaranteed convergence and error (in the least-squares sense).

In this paper, we examine the properties of an alternate decomposition, called semi-discrete transform (SDD). Unlike SVD, SDD terms only require $\log_2 3$ bits per entry (since entries can take values -1/0/1) and a floating point number for each scalar corresponding to the singular value. This enables approximation of the matrix to much higher ranks for the same amount of storage. We show that this results in significant improvements in image quality over SVD for the same compression ratios. We also show that the image quality is still inferior to discrete cosine transform (DCT) coded images that are commonly used in image and video compression. However, decoding SDD coded images is much faster than computing inverst DCTs. Furthermore, SDDs facilitate a wide range of optimizations such as pattern matching and motion compensation, which can further improve compression significantly.

16.2 THEORETICAL UNDERPINNINGS

In this section, we briefly overview semi-discrete matrix transforms (SDD) [5] and their application to image compression. A k-term SDD of an $m \times n$ matrix A is a decomposition of the form

$$
A_k = \underbrace{[X_1 X_2 \ldots X_k]}_{X} \underbrace{\begin{bmatrix} d_1 & 0 & \ldots & 0 \\ 0 & d_2 & \ldots & 0 \\ \vdots & \vdots & \ddots & \vdots \\ 0 & 0 & \ldots & d_k \end{bmatrix}}_{D_k} \underbrace{\begin{bmatrix} Y_1^T \\ Y_2^T \\ \vdots \\ Y_k^T \end{bmatrix}}_{Y^T}
$$

$$
= \sum_{i=1}^{k} d_i X_i Y_i^T.
$$

Here, each X_i is an m-vector with entries from the set $S = \{-1, 0, 1\}$, each Y_i is an n-vector with entries from the set S, and each d_i is a positive scalar. Notice that this differs from conventional singular value decomposition where the set S is the set of all reals. We call this a k-term SDD. Although every matrix can be expressed as an mn-term SDD

$$A = \sum_{i=1}^{m} \sum_{j=1}^{n} a_{ij} e_i e_j^T,$$

where e_k is the k-th unit vector. The usefulness of the SDD is in developing approximations that have far fewer terms. Consequently, a suitable truncated k-term approximation is used. Let A_k denote the k-term approximation ($A_0 \equiv 0$). Let R_k be the *residual* at the kth step; that is $R_k = A - A_{k-1}$. Then the optimal choice of the next triplet (d_k, X_k, Y_k) is the solution to the subproblem

$$\min_{X,Y,d} \|R_k - d_k X_k Y_k^T\|_F^2$$

$$X \in S^m, Y \in S^n, d > 0,$$

where $\| \cdot \|$ represents the Frobenius norm. Restricting vectors X and Y to the set S has the obvious advantage of requiring only $\log_2 3$ bits per vector element as opposed to a much higher number of bits in the case of SVD. However, the above optimization problem has been shown to be NP-complete. Kolda and O'Leary [4] show that excellent heuristics exist for approximating the solution to this optimization problem. Furthermore, since our application does not require a global minimum, a heuristic solution is found to be computationally efficient and adequate.

16.3 COMPARISON OF MATRIX TRANSFORMS FOR IMAGE COMPRESSION

We examine SVD and SDDs in the context of two distinct image compression algorithms. The first one, called the *fixed-cell size algorithm* is similar to the kernel of JPEG and MPEG schemes. This algorithm divides the image into rectangular cells. The SDD of each cell is processed independently. A maximum acceptable error is set as a parameter, and for each cell, the algorithm computes as many terms as necessary to satisfy the error bound[1]. In order to ensure that the compression ratio does not deteriorate because of the noise in images that are discontinuous and non-uniform (e.g., the feather ornament that hangs from Lena's hat), we also specify a maximal SDD level.

In the second algorithm, which we call the *adaptive cell size algorithm*, the rank of the SDD and the maximum acceptable error are fixed. This algorithm uses a hierarchical approach to recursively subdivide the image into smaller cells. For each cell, we compute an SDD approximation. If the difference between the approximation and the original matrix is larger than the specified error, the image is split into four subimages, and the process is repeated on the *difference* between the SDD approximation and the original image for each subimage. This is repeated until the image can be approximated or until the subimages reach a specified minimum threshold size. Every new term in this procedure can be interpreted as a correction, or a foreground that is superposed on the previous terms to reconstitute the original image. Figure 16.1 illustrates a representation of the recursive partition for the image Lena.

---

[1] We choose the Frobenius, norm, but similar results were obtained with the infinite norm.

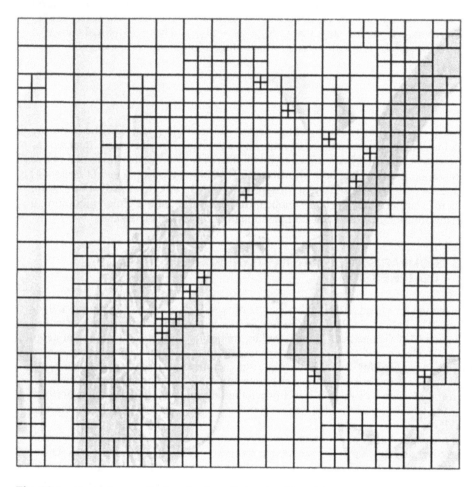

Fig. 16.1 Clustering map for the adaptive cell size algorithm. The corresponding compressed image has been added in the background in light colors.

(a)

(b)

Fig. 16.2 Comparison of SVD (a) and SDD (b) encoding techniques with fixed cell size 16×16 pixels at 1 bpp.

253

Based on these two algorithms, we experiment with a variety of variants to exhaustively compare the performance of SVD, SDD, and DCT-based compression. A comparison of SVD and SDD-based compression for fixed cell size is presented in Figure 16.2. The figure clearly illustrates that at compression rates of 1 BPP, SDD is considerably better than SVD. This result holds across all images and compression rates we have experimented with. We also use a variable cell-size (varying from 8×8 to 32×32 pixels) along with truncated SVD, SDD, and DCT to encode the cells. Figure 16.3 illustrates the results of this experiment for Lena. It is evident from this figure that SDD yields images of significantly better quality than SVD. It is also clear that DCT outperforms both of these transforms. However, as we shall see in the next section, SDD provides a convenient framework that can be effectively used for pattern matching within and across frames.

16.4 PATTERN MATCHING USING SDDS

Pattern matching [1, 2], in this context, refers to the process of replacing some occurrences of a pattern (a cell) by a pointer to this pattern, and, if necessary, its dimensions. Pattern matching is the most time-consuming and efficient storage mechanism used in most conventional video compression schemes. Most often, when two cells match, they are in the same neighborhood. In this case, pointers are stored as offsets from current position as opposed to global pointers- for example, restricting matches to neighbors requires only 2 bits to determine if a cell is be stored as a pointer to its left or up neighbor. For Lena, using the same distortion parameters as the tests reported in Table 16.1, 130 cells could be replaced by a local pointer towards either the left or up-neighbor (as opposed to 275 using a global pattern match).

16.4.1 Accelerating Search Using SDDs

A simple pattern matching algorithm builds a *growing database* of all unique patterns that have been previously encountered. Each cell is compared to the elements of the database until a match is found, or until the end of the database is reached. In the latter case, the cell is added to the database. Two cells "match" if the norm of their difference is smaller that the maximal acceptable error. Only the database and the pointers are stored. Such an algorithm is particularly costly in terms of CPU time for two reasons:

- It requires a large number of comparisons since every cell must be compared to the elements of the database, until a match is found.

- The process of comparing two cells requires computation of a norm of difference between the cells.

A major part of this computation can be avoided, using the fact that for an image, in most cases, the norm of cell difference is large and can be eliminated by pruning heuristics. SDD coefficients that have been previously computed can be effectively used to prune the search space for possible pattern matches. Cells that are not eliminated are potential matches, and are compared explicitly. We have developed two pruning heuristics for the fixed cell and adaptive cell size algorithms.

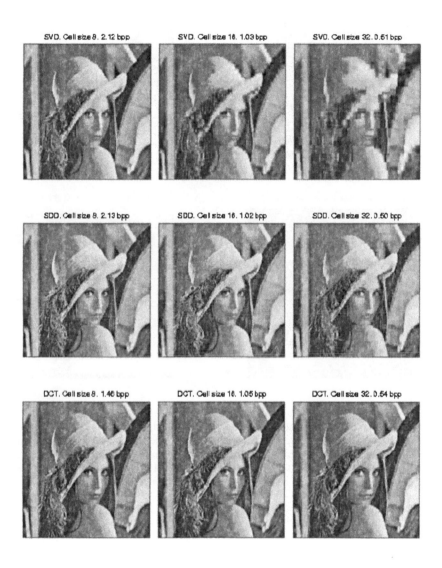

Fig. 16.3 Comparison of image compression techniques: SVD (1st row), SDD (2nd row), and DCT (3rd row). The first column corresponds to images at 2bpp (except for DCT), the second column at 1bpp, and the third column at 0.5 bpp.

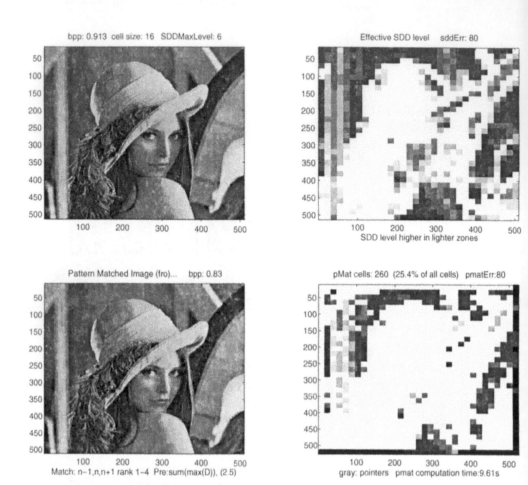

Fig. 16.4 Use of SDD for pattern matching: figure top-right shows the required SDD level for each cell (of size 16) with darker shades indicating higher SDD level. Figure bottom-right shows improvements by pattern matching SDD blocks. In this matching schemes, we only compute matches with respect to other blocks at SDD level one below or one above the SDD level of the current block. The result is an extremely fast matching algorithm that runs in 9.6 seconds as a matlab script on a 400 MHz Pentium II.

Heuristics	Time	Cells
Level 0	100	279
Level 1	61	276
Level 2	47	245
Level 3	61	273
Level 4	8.9	274

Table 16.1 Improvements in pattern-matching search obtained by pruning the search. Table illustrates search time (Matlab scripts on Sparc Ultra I) and number of cells matcles. Various optimization levels are as follows: Level 0: No optimizations (Basic pattern-matching algorithm); Level 1: Only matches SDD levels 1 to 4; Level 2: Matches within same SDD level; Level 3: Matches within same SDD level and level ±1; and Level 4: Within SDD levels 1 through 4, matches the same SDD level as well as level ±1.

Pruning method for the fixed cell size algorithm. The SDD level of a given cell depends on the intrinsic complexity of the cell. It is unlikely that two cells that contain complex patterns (i.e, for which a large SDD level was required to be within the error bound) will match. As a matter of fact, we notice, for example that for Lena, using a 16×16 cell decomposition and a maximal SDD level of 10, all the matches were between cells that had an SDD level betwccn 1 and 4. We can therefore restrict cell comparisons to cells for which SDD levels are comparable. Hence, instead of constructing a single database with all previously encountered patterns, we make as many small databases as the numbers of SDD levels that are relevant for pattern-matching (4 databases in our example). Table 16.1 shows that comparing a given pattern with all the patterns from its database (e.g., with all the patterns that have the same SDD decomposition), and with its immediate neighbors (patterns from thc SDD Level ±1) is a good compromise. This heuristic provides very similar results as comparing an element with the entries from all databases (in terms of number of matches) while yielding over an order of magnitude increase in speed.

The next step in pruning the search is accomplished by preprocessing the cells that have not been already eliminated for comparison. Here again, the idea is to eliminate as many candidates for pattern-matching as possible, with a fast test. Typically, a cell that contains high values will not match a cell that contains low values (under the Frobenius norm that was used in these experiments. Similar heuristics can be constructed for other norms as well). Such information is available in the scalar values, d_i, of the SDD. Using the difference of these singular values, we can quickly prune possible matches further. Finally, we perform the basic pattern matching algorithms on the remaining cells.

Pruning method for the adaptive cell size algorithm. In the previous section, we used SDD to evaluate a bound on cells that had the same size. We can take advantage of the tree structure in the adaptive cell algorithm to determine in one operation if a given pattern (cell) matches any of the submatrices of a bigger cell. Here again, the search is accelerated by preprocessing using the singular values. The remaining cells are potential matches, and the actual matches are determined using the basic growing database pattern-matching algorithm.

16.4.2 Computing a Motion Vector for Video Encoding

Typical video encoding schemes rely on identifying zones that have been translated from neighboring frames. This translation is often called a *motion vector*. We have developed an algorithm using SDDs to compute motion vectors efficiently. This implementation can identify patterns that have moved by an integral number of cells. Our approach is summarized in the following steps for computing the motion vectors between two images:

- Compute the SDD of the first frame.

- Identify the cells in the current image that are different from the original frame.

- Compute the SDD of the all the cells from the "difference index", generated in the previous step. Among these, there are patterns that correspond to a new position of the foreground object, and some that correspond to the background that is now visible, because the object that was there on the original image has moved.

- Using the SDD values that have been computed find the patterns in the current image that correspond to a pattern in the previous image and create a motion vector.

- Redraw the background.

Figure 16.5 shows a few motion vectors for patterns that were inserted on top of a still background.

16.5 CONCLUDING REMARKS

In this paper, we have shown that discrete transforms such as SDD are superior to their continuous counterparts, namely SVD. We have shown that DCTs are still superior to SDDs. However, we have shown that SDDs are extremely useful in computing pattern matches and motion vectors in video coding. Furthermore, decoding SDD-coded images only requires logical operations and additions. This makes them extremely fast. Our current work in this area focuses on collecting detailed experimental results and building a robust compression framework upon the SDD formalism.

16.6 ACKNOWLEDGEMENTS

This work was supported by NSF Grants C-CR-9804760, EIA-9806741, ACI-9875899, and ACI-9872101, and contracts from sponsors of CERIAS at Purdue. Computing equipment used for this work is supported by the Intel Corp.

Original Frame   0.902 bpp. (NO PMAT)

Current Frame   0.0146 bpp.

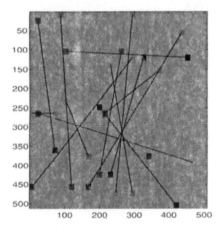

Fig. 16.5 Using the SDD based motion vector algorithm, 14 out of 15 patterns were identified. In this experiment, the cell size is 16×16, maximal SDD Level is 6, original SDD computation time is 55.9s, additional SDD and motion vector computing time is 3.71s (All times using Matlab scripts on a Sparc Ultra I).

REFERENCES

1. M. Alzina, W. Szpankowski, A. Grama. 2D-Pattern Matching Image and Video Compression: Theory, Algorithms, and Experiments. *IEEE Transactions on Image Processing*, 11, 318–331, 2002.

259

2. M. Atallah, Y. Génin, and W. Szpankowski. Pattern Matching Image Compression: Algorithmic and Empirical Results. *IEEE Transactions on Pattern Analysis and Machine Intelligence* vol.21, no 7, 1999.

3. Gene H. Golub and Charles Van Loan. *Matrix Computations*, Second Edition. The Johns Hopkins University Press, Baltimore, MD, 1989.

4. T. G. Kolda and D. P. O'Leary. Computation and Uses of the Semidiscrete Matrix Decomposition. *ACM Transactions on Information Processing*, 1999.

5. D. O'Leary and S. Peleg. Digital Image Compression by Outer Product Expansion. *IEEE Transaction on Communications*, 31, 441-444, 1983.

6. M. Rabbani and P. Jones. *Digital Image Compression Techniques.* Bellingham,Wash. :SPIE Optical Engineering Press, 1991.

Author(s) affiliation:

- **Sacha Zyto, Ananth Grama, and Wojciech Szpankowski**

  Department of Computer Sciences
  Purdue University, W. Lafayette, IN 47907, USA.
  Email: [zyto, ayg, spa]@cs.purdue.edu

17

Internet Security: Report on ICANN's Initiatives and on the Discussions in the European Union

Stefano Trumpy

17.1 SECURITY ASPECTS

The Internet is designed as a highly distributed system, with very few single points of control or failure. The computers that interconnect all the networks world-wide are consequently very numerous and widely distributed globally. They are also owned and controlled by many distinct organizations and individuals.

Consequently, general co-ordination of the Internet, and a fortiori security, depends on widespread co-operation.

One major objective of the distributed architecture of the Internet is precisely that of insulating the networks as a whole from any particular failure. Indeed, as seen, the Internet responded to the 9/11 events very robustly, particularly in the United States and Europe, the loss of a major telecommunications facility adjoining the WTC notwithstanding. But this may not have been the case in more peripheral parts of the Internet; South Africa reported a significant and prolonged outage at the time.

However, certain Internet functions are centralized due to the necessity of unique assignments of names, addresses and protocols. These are the functions that are under the auspices of ICANN itself. Accordingly, should there be vulnerabilities and failures in the future, they might occur in these particular areas.

17.2 RETHINKING SECURITY AFTER SEPTEMBER 11TH 2001: WHICH ROLE FOR ICANN?

One by-product of the relative success of the Internet under stress has been a certain sense of complacency among the operators of central Internet functions. Their preferred philosophy is to proceed on the basis of informal co-operation and best practice guidelines for the operation of the Internet infrastructure, such as RFC 2870. However, responsibility for monitoring and enforcing compliance is not sufficiently clear at present and the respective roles of ICANN, governments and/or external auditing need to be clarified.

In the recent ICANN meetings in Marina del Rey (November 10-15th) a wide range of potential problems and issues where discussed; a detailed analysis of the outcome will be prepared by ICANN staff and other participants. The following points were addressed in particular and ICANN can be expected to focus on these matters.

In general, since security and reliability are relative concepts, the consensus was to approach the question systematically through risk analysis and auditing performance with respect to agreed best practices, rather than setting formal, contractual requirements. In any event, absolute security is an impossibility. What is required is to balance the costs and inconvenience of heightened security with the requirements of reliability and confidence.

17.3 DISCUSSION OF SECURITY MEASURES TO BE ADOPTED BY ICANN

During the third annual meeting of ICANN the following arguments were discussed:

1. Vulnerabilities

    (a) DDOS (Distributed Denial of Service)

    Certain vulnerabilities are dependent on general problems in the networks. For example, distributed denial of service (DDOS) attacks will continue to be a significant threat in the future. In this context, the Chairman of RIPE drew attention to a large portion of mass-produced personal computers that can be connected to the Internet as soon as they are purchased. They are unwary sitting targets for the perpetrators of DDOS attacks. This issue should be taken up with PC vendors. It is not clear whether the Root Server System or the DNS name servers are particularly likely to be a target for DDOS. To date, other more visible large commercial sites have been targeted.

    (b) Complexity

    Systems become more vulnerable as their complexity and inter-dependencies increase. This is relevant to the Internet. Modern software packages are much more complex than they were a few years ago. In practice, vulnerabilities have been inadvertently increased by their use. It is probably impossible to avoid bugs and other weaknesses in such systems while correcting errors through "patches" is now getting out of hand. Increasing standardisation on a few operating systems and major software packages increases overall vulnerability. Lack of diversity is another source of risk. In particular, the software running the Internet centralised functions should be diverse and recoverable from secondary or back-up sites.

(c) Root Server System

Currently the Root Server System is still managed by volunteer organisations and individuals (generally unpaid), among whom are some of the best and most dedicated engineers on the Net, working in close co-operation with each other. These activities are hosted by several different organisations, including the US Department of Defence and other US Government departments. However, commercial users and the ccTLD Registries are uncomfortable with the informality of the current regime, and said that they would expect such a critical infrastructure function to be the subject of a more rigorous and transparent regime of service contracts. The debate was inconclusive, but there appeared to be a consensus that, in due course, the second generation of Root Server operators should come under a more formal regime.

From the security point of view, the Root Server System currently contains sufficient redundancies to be able to shrug off and survive an attack limited to any one (or a few) of the 13 Root Servers. On the other hand, should all 13 Root Servers be the target of a comprehensive DDOS attack, the situation could temporarily become very serious. The consensus in the meetings was that this would be highly unlikely. However, the possible risk has led one of the US ICANN Board members (Karl Auerbach) to advocate systematic mirroring of the Root Servers and public availability of all zone file data.

Most commentators agree that too many of the Root Servers (ten out of thirteen) are in the United States. But there are engineering constraints on either increasing the number of root servers or on re-locating them in the short term. Physical re-location of a Root Server (for example to a different country) would involve changing its IP address. This would require updating the corresponding information currently encoded in hundreds of thousands of name-servers, world-wide. Increasing the number of Root Servers would require going back to IETF to modify the protocol.

ICANN will issue a report on the Root Server system before the end of 2001. The current philosophy is to develop a MOU among ICANN and the 13 Root Servers and encourage implementation of best practice guidelines based on RFC 2870: "Root Name Server Operational Requirements" dated June 2000. Some commentators argued that RFC 2870 should now be revisited, in view of the increased attention paid to security aspects.

(d) DNS Name Servers

The name servers replicate all or part of the information derived from the Root Servers and the TLD Registries, that is necessary for routing IP packets throughout the Internet. Following the recommendations of RFC2182, each TLD should have at least two secondary servers and possibly more (in practice, the number varies from 2 to 8), located in different autonomous systems. Most ISPs will cache (copy) some of this data to accelerate resolution of enquiries by their customers. Consequently there are very many name-servers in the DNS. There would appear to be several problem areas here:

263

- Significant numbers of ccTLD Name Servers also act as secondary servers for several TLDs. Thus secondary servers of other TLDs are maintained on the same TLD server. It is not clear whether this situation constitutes an additional risk of propagation in the event of a virus or a DDOS attack. Furthermore, numerous TLD Name Servers operate obsolete versions of the BIND software, some of which are known to be insecure.

- The underlying quality of registration data from which the Whois data and the Zone Files are derived, leaves a great deal to be desired. Some technical reports suggest that as many as two-thirds of the domain names have erroneous ("mis- configured") data, possibly arising from lax maintenance by Registrars and Registries and from the speculative character of a significant proportion of domain name registrations. The Internet continues to function, these shortcomings notwithstanding. However it is difficult to argue that DNS Registration data is relevant to identify perpetrators of DDOS, or eventually for law enforcement, while at the same time condoning such poorly recorded data.

- For some time a new more secure DNS protocol known as DNS-Sec, has been under development. Although DNS-Sec is already implemented and functioning, the new product implies an increase of an order of magnitude in data files size, thus imposing remarkable operational overheads. Consequently, DNS-Sec is not yet deployed, and the market does not appear to support it spontaneously. However, the new system will be necessary for building the chain of trust necessary to develop electronic commerce. New versions of the system aimed at reducing these complexities have been developed. More generally, this is perhaps an illustration of the question of how the costs of more secure operation in the Internet will be allocated and paid for. In any event, DNS-Sec would facilitate authentication of DNS registration, modifications and queries. It would not address other identified weaknesses in the DNS.

2. Action to be taken

What has to be considered now is the future security of the Internet in any possible critical situation. The users, the market and the governments are demanding this. The studies to be undertaken are likely to include:

- risk analysis
- locating single points of failure
- disaster recovery
- best practices for operators

The roles of ICANN and other Internet participants will need to be defined. ICANN's Chief Executive, Stuart Lynn, argued that ICANN should operate through incentives, transparency and encouraging best practices rather than through contractual obligations and sanctions. In any event, a significant work-programme is anticipated in the immediate future, coming from the recently created "Standing Committee on Stability and Security."

3. Position of the Governmental Advisory Committee

The Chair of the GAC, Paul Twomey, stated that, in the event of serious problems with the Internet infrastructure, the focus - and the blame - would be on governments, and not on ICANN or the Internet operators. Governments have a direct interest in the quality of the operation, the results of risk analysis and the nature of the improvements proposed.

Arguably, national governments will be more sensitive to the security aspects of the DNS than in the case of other aspects of Internet management, due to their broader responsibilities for network security. They also have a clear interest in the ccTLD Registries operating within their jurisdictions, whereas ICANN might be expected to take the lead with the Root Servers and the generic Top Level Domains.

However, the overall picture is far from straightforward. Several countries' ccTLDs are operated from other locations, including in the US and, in a few cases, the EU. Many ccTLD name servers are naturally located in other parts of the world. Establishing a chain of responsibility and accountability in these circumstances may be technically, if not politically, complex. The level of public responsibility for the many name servers operated by the ISP industry is also moot.

The consensus in the GAC was that governments should emphasise mutual responsibility rather than sovereignty and unilateral authority in these areas.

17.4 SECURITY MEASURES TAKEN BY THE EUROPEAN UNION

The Council of the European Union - Telecommunications Working Party is discussing a draft Council Resolution on a common approach and specific actions in the area of network and information security. In this section some of the main areas of concern and proposed actions are presented.

The resolution to be adopted is a comprehensive position, which rests on a number of previous recommendations and directives adopted in the recent past. The ongoing discussion is based on the following prerequisites:

1. networks and communications systems have become a key factor in economic and social development and their availability and integrity are crucial to essential infrastructures, as well as to most public and private services and the economy as a whole;

2. in the light of the increasingly important role played by electronic services, the security of networks and information systems is of growing public interest;

3. the complex nature of the network and information security means that in developing policy measures in this field, public authorities must take into account a range of political, economic, organisational and technical aspects, and be aware of the decentralised and global character of communication networks;

4. the Internet infrastructure should provide the greatest possible availability and be managed and operated in a robust and transparent manner;

5. network and information security means:

- ensuring the availability of services and data
- preventing the disruption and unauthorized interception of communications
- confirmation that data which have been sent, received or stored are complete and unchanged
- securing the confidentiality of data
- protection of information systems against unauthorized access
- protecting against attacks involving malicious software and securing dependable authentication

The Member States are asked:

- to launch or strengthen information and education campaigns to increase awareness of network and information security;

- to promote best practices in security policies based, where appropriate, on internationally recognized standards;

- to review national arrangements regarding computer emergency response, taking the necessary action as regards their ability to prevent, detect and react efficiently at national and international levels against network and information system disruption and attack;

The Commission is invited:

- to facilitate an exchange of best practice regarding awareness-raising actions and to draw up an initial inventory of the various national information campaigns;

- to reinforce the EU dialogue and co-operation with international organisations and partners on network security, in particular on the implications of the increasing dependency on electronic communication networks;

- to propose a structure for a more open and transparent operation of critical parts of the Internet infrastructure, including the root server system;

- to set up a " Cyber - Security task force"

17.5 BILATERAL CONSULTATION OF EU WITH THE USA ON ICANN'S ACTIVITY

Several aspects of ICANN's meetings should be included in the agenda of the continuing bilateral dialogue between the EU and the USA on Internet management policies. These could include:

1. Confirmation that Internet Security is a joint, co-operative task for the private sector and governments. It is not technically possible for the US or the EU (or other governments) to achieve the necessary results unilaterally. The EU will address these issues for the critical facilities for which we are responsible, in

266

conjunction with the Member States and the private sector, but the practical outcome will require considerable co-operation with the USA, other governments and with ICANN.

2. Moving towards a more balanced and transparent Internet management system, including a clarification of ICANN's role and responsibilities. Achieving a more balanced geographical distribution of the current 13 Root Servers, taking full account of current and future Internet traffic flows. The EU is looking forward, with great anticipation, to the actions that ICANN will take on the basis of the forthcoming report on the Root Server System.

3. More generally, we expect ICANN to initiate an action programme to follow up on the results of the MdR meetings. The EU will keep in close touch with further work in this area, directly with ICANN and through the Governmental Advisory Committee (GAC). Elements of this programme may be discussed with the USA bilaterally, should the need arise.

17.6 MIGRATING FROM AN INTERNET BASED ON BEST EFFORT TOWARDS A NETWORK WITH GUARANTEED LEVEL OF PERFORMANCE

This is a very complex question. Internet was born and subsequently has grown up generally based on best effort criteria. The routing is assured by a best effort approach; the services on the network are assured on the same criteria, based on the availability of the end to end bandwidth. E-commerce and professional services need to foresee an evolution of the network towards a globally guaranteed level of services/ availability. This will imply higher costs and the co-operation of many actors for which the quality of service is essential in their operations. This will imply a remarkable effort to define new organizational models. Also the quality of service will have to be assured by new contract-based conditions involving the key partners. Perhaps a best effort network at lower costs will survive for those users with simpler needs or who are unable to afford the costs of the network with guaranteed level of performance.

Author(s) affiliation:

- **Stefano Trumpy**

  Research manager at CNR/IAT
  Responsible of international relations of ".it" registry
  Italian delegate in the Governmental Advisory Committee of ICANN
  Via G. Moruzzi 1
  56100 Pisa, Italy
  Email: stefano.trumpy@iat.cnr.it

List of Workshop Attendees

Professor **Gul A. Agha**

    University of Illinois at Urbana-Champaign
    Department of Computer Science
    W Springfield Ave
    Urbana
    Illinois 61801, USA
    Phone: +1 (217) 244-3087
    FAX: +1 (217) 244-6869 1304
    Email: agha@cs.uiuc.edu
    http://www-osl.cs.uiuc.edu

Dr. **Jeffrey M. Bradshaw**

    Research Scientist
    Institute for Human and Machine Cognition
    University of West Florida
    40 South Alcaniz Pensacola
    Florida 32501, USA
    Phone: +1 (850) 202-4400 (main IHMC number)
    Fax: +1 (850) 202-4440
    Email: jbradshaw@ai.uwf.edu
    http://www.coginst.uwf.edu/~jbradsha/

Dr. **Frederica Darema**

    Senior Science and Technology Advisor
    National Science Foundation

4201 Wilson Boulevard
Arlington
Virginia 22230, USA
Room:1160 N
Phone: +1 (703) 292-8980
Fax: +1 (703) 292-9030
Email: fdarema@nsf.gov

Professor **Renato Figueiredo**

Department of Electrical and Computer Engineering
Northwestern University
2145 Sheridan Road
Evanston
Illinois 60208-3118, USA
Phone: +1 (847) 491-5410
Fax: +1 (847) 491-4455
Email: renato@ece.nwu.edu

Professor **José A.B. Fortes**

Professor and BellSouth Eminent Scholar
Director, Advanced Computing and
          Information Systems (ACIS) Laboratory
Department of Electrical and Computer Engineering
University of Florida
P.O. Box 116200, 339 Larsen Hall,
Gainesville
Florida 32611-6200, USA
Phone: +1 (352) 392-9265
Fax: +1 (352) 392-5040
E-mail: fortes@ufl.edu
http://www.fortes.ece.ufl.edu

Professor **Erol Gelenbe**

Director School of EECS
Associate Dean of Engineering and Computer Science
University of Central Florida
Orlando
Florida 32816, USA
Phone: +1 (407) 823 0345
Fax: +1 (407) 823 5419
Email: erol@cs.ucf.edu

Dr. **Philippe Jacquet**

Responsable Scientifique du Projet HIPERCOM
INRIA, Unité de recherche de Rocquencourt
Domaine de Voluceau
Rocquencourt - B.P. 105 78153
Le Chesnay Cedex - France

Phone: +33   (1) 39 63 52 63
Secrétary: +33   (1) 39 63 53 63
Email: Philippe.Jacquet@inria.fr

Dr. **Yongchang Ji**

School of Electrical Engineering and Computer Science
University of Central Florida
Orlando
Florida 32816, USA
Phone: +1 (407) 823 4860
Fax: +1 (407) 823 5419
Email: yji@cs.ucf.edu
http://www.cs.ucf.edu/~yji

Dr. **Craig A. Lee**

Section Manager, High Performance Computing
The Aerospace Corp.   M1-102
2350 East El Segundo Blvd.
El Segundo
California 90245, USA
Phone: +1 (310) 336-1381
Fax: +1 (310) 336-4402
Email: lee@aero.org
http://www.aero.org

Professor **Dan C. Marinescu**

School of Electrical Engineering and Computer Science
University of Central Florida
Orlando
Florida 32816, USA
Phone: +1 (407) 823 4860
Fax: +1 (407) 823 5419
Email: dcm@cs.ucf.edu
http://www.cs.ucf.edu/~dcm

Professor **Andrea Omicini**

DEIS, Università degli Studi di Bologna,
Sede di Cesena
Via Rasi e Spinelli 176
47023 Cesena, FC, Italy
Phone: ++ 39 0547 614552
Fax: ++ 39 0547 614550
Email:aomicini@deis.unibo.it
http://lia.deis.unibo.it/~ao

Professor **Gian Pietro Picco**

Dipartimento di Elettronica e Informazione
Politecnico di Milano

271

Piazza Leonardo da Vinci, 32
20133 Milano, Italy
Phone: +39-02-2399-3519
Fax: +39-02-2399-3411
Email: picco@elet.polimi.it

Dr. Thierry Priol

IRISA
Campus Universitaire de Beaulieu
35042 Rennes Cedex, France
Phone: +33 2 99 84 72 10
Secretary: +33 2 99 84 72 28
Email: Thierry.Priol@inria.fr

Dr. Davide Rossi

Dipartimento di Scienze dell'Informazione
Università degli Studi di Bologna
Mura Anteo Zamboni, 7
Bologna, Italy
Phone: +39-051 2094871
Fax   +39-051 2094510
Email: rossi@cs.unibo.it

Dr. Niranjan Suri

Research Scientist,
Institute for Human and Machine Cognition
University of West Florida
40 South Alcaniz Pensacola
Florida 32501, USA
Phone: +1 (850) 202-4444
Fax: +1 (850) 202-4440
Email: nsuri@uwf.edu

Professor Wojciech Szpankowski

Computer Science Department
1398 Computer Science Building
West Lafayette,
Indiana 47907-1398, USA
Office Phone: +1 (765) 494-6703
FAX: +1 (765) 494-0739
Email: spa@cs.purdue.edu

Dr.-Ing Robert Tolksdorf

Technische Universität Berlin,
Fachbereich 13, Informatik,
Formale Methoden, Logik und Programmierung (FLP),
Room FR6071
Franklinstraße 28/29,

D-10587 Berlin,
Phone: +49-30-314-25184,
Fax: -73622,
Secretary: -73540,
Email: research@robert-tolksdorf.de

Professor **Anand Tripathi**

Department of Computer Science & Engineering
University of Minnesota
Minneapolis
Minnesota 55455, USA
Phone: +1 (612) 625-9515
Fax: +1 (612) 625-0572
Email: tripathi@cs.umn.edu

Dr. **Stefano Trumpy**

Research Manager of CNR
National Research Council
Via G. Moruzzi 1
56100 Pisa, Italy
Email: Stefano.Trumpy@iat.cnr.it
Phone: +39-050-3152634,
Fax: +39-050-3152593

Professor **Lotfi A. Zadeh**

Computer Science Division
University of California
Berkeley
California 94720-1776, USA
Phone: +1 (510) 642-4959
Fax: +1 (510) 642-1712
Email: zadeh@cs.berkeley.edu

D. 10587 Berlin
Phone: +49-30/3 14 26 84
Fax:
Sekretariat:
Email:

Milton Keynes UK
Ingram Content Group UK Ltd.
UKHW031145141024
449569UK00024B/1054